# Palgrave Studies in Sustainability, Environment and Macroeconomics

Series Editor
Ioana Negru
Department of Economics
SOAS, University of London
London, UK

Most macroeconomic theory and policy is orientated towards promoting economic growth without due consideration to natural resources, sustainable development or gender issues. Meanwhile, most economists consider environmental issues predominantly from a microeconomic perspective. This series is a novel and original attempt to bridge these two major gaps and pose questions such as: Is growth and sustainability compatible? Are there limits to growth? What kind of macroeconomic theories and policy are needed to green the economy?

Moving beyond the limits of the stock-flow consistent model, the series will contribute to understanding analytical and practical alternatives to the capitalist economy especially under the umbrella term of "degrowth". It will aim to reflect the diversity of the degrowth literature, opening up conceptual frameworks of economic alternatives — including feminist political ecology — as critical assessments of the capitalist growth economy from an interdisciplinary, pluricultural perspective.

The series invites monographs that take critical and holistic views of sustainability by exploring new grounds that bring together progressive political economists, on one hand, and ecological economists, on the other. It brings in.

More information about this series at
http://www.palgrave.com/gp/series/15612

Michael Roos • Franziska M. Hoffart

# Climate Economics

A Call for More Pluralism And Responsibility

Michael Roos
Faculty of Management and
Economics
Ruhr University Bochum
Bochum, Germany

Franziska M. Hoffart
Faculty of Management and
Economics
Ruhr University Bochum
Bochum, Germany

ISSN 2635-2621          ISSN 2635-263X   (electronic)
Palgrave Studies in Sustainability, Environment and Macroeconomics
ISBN 978-3-030-48422-4        ISBN 978-3-030-48423-1   (eBook)
https://doi.org/10.1007/978-3-030-48423-1

This Palgrave Macmillan imprint is published by the registered company Springer Nature
Switzerland AG.
The registered company address is: Gewerbestrasse 11, 6330 Cham, Switzerland

# Acknowledgement

We would like to express our great appreciation to Klaus Steigleder for his valuable feedback on early drafts. As Michael's esteemed colleague and Franziska's former supervisor, he played an important role in sparking our deep interest in reflecting on the ethics of climate change. Our special thanks are extended to the colleagues at the Institute for Macroeconomics. We appreciate the fruitful discussions and their support with proofreading and the collection of data.

I wish to thank my parents, Brigitte and Rainer, for their love and unconditional support throughout my life and studies. Special thanks also to my friends, especially Nora and Hanna, who stand by me during every struggle and provide inspiration in my life. (Franziska M. Hoffart)

I would like to thank my wife and children for their patience during yet another writing episode and many hours at the computer. I am also grateful for their willingness to listen to my lectures over breakfast and dinner. (Michael Roos)

# CONTENTS

# LIST OF FIGURES

# LIST OF TABLES

# CHAPTER 1

# Introduction

**Abstract** In the introduction, the authors present the book's main arguments. Roos and Hoffart argue that climate change is an economic problem, but does not receive enough attention in economic mainstream research. Furthermore, the policy recommendations derived from mainstream research are biased towards achieving efficiency rather than effectiveness. This emphasis on efficiency implies a value judgment that is not discussed enough. The authors argue that economics should contribute to the transformation of society towards more sustainability in a more practical way than it currently does. More pluralism of scientific approaches would help to make economics more transformative. At the end of the introduction, the following chapters are briefly summarised.

**Keywords** Climate change • Optimal climate policy • Mitigation • Precautionary principle • Transformative science

## 1.1 CLIMATE CHANGE

Climate change is the biggest challenge mankind currently faces. The two hot summers of 2018 and 2019, with heat waves in Europe, made everybody experience at last what climate change might mean in the future.

© The Author(s) 2021
M. Roos, F. M. Hoffart, *Climate Economics*, Palgrave Studies in
Sustainability, Environment and Macroeconomics,
https://doi.org/10.1007/978-3-030-48423-1_1

Global warming will affect the living conditions in many countries severely, but especially in the developing world. Apart from extreme heat, water scarcity and negative impacts on food production are among the most alarming consequences. Regional and international migration triggered by climate change will also affect richer countries, in which the direct effects of climate change might be more limited. The Swedish girl Greta Thunberg gave climate activism a face and sparked the *Fridays for Future movement* of young protesters, which generated much public awareness that a policy change is needed. These protests are not just mass hysteria. In the German-speaking countries, almost **27,000** scientists supported the concerns of the young protesters and declared: "these concerns are justified and are supported by the best available science. The current measures for protecting the climate, biodiversity, and forest, marine, and soil resources, are far from sufficient".[1] In contrast to scepticism spread by organised climate change denial (Oreskes and Conway 2010; Dunlap and McCright 2011), there is a vast scientific consensus that anthropogenic climate change[2] exists. Today there is accumulating evidence that it is occurring significantly faster than expected.

In a special report published in October 2018, the IPCC estimated that the global average temperature is already approximately 1.0 °C above pre-industrial levels mainly due to human activities (IPCC 2018). In the Paris Agreement of 2015, 197 parties signed to undertake measures to keep the global temperature increase below 1.5 °C. According to the IPCC (2018), it is likely that global warming will reach 1.5 °C between 2030 and 2052, if the current rate of temperature increase is unchanged. However, new scientific evidence suggests that climate change unfolds much faster than anticipated so far. Xu et al. (2018) expect that the 1.5 °C level could be reached already in 2030, because the combined effect of rising $CO_2$ emissions, declining air pollution and natural climate cycles has been underestimated. Cheng et al. (2019) report that the oceans warm up 40% faster than previously assumed by the IPCC. Evidence is also mounting that permafrost thaw in the Arctic happens much faster than previously predicted. As a result highly aggressive greenhouse gases such as methane and $N_2O$ are emitted at higher rates than previously thought (Schoolmeester et al. 2019; Wilkerson et al. 2019; Farquharson et al. 2019). This is one of

[1] https://www.scientists4future.org/stellungnahme/statement-text/.
[2] https://climate.nasa.gov/scientific-consensus/, https://skepticalscience.com/global-warming-scientific-consensus-basic.htm.

the vicious cycles described by climate science, namely that global warming due to greenhouse gas emissions sets free further greenhouse gases so far bound in permafrost. Furthermore, large glaciers worldwide and even the ice mass in Greenland seem to melt much quicker than predicted (Bevis et al. 2019; Zemp et al. 2019). The melting of large ice masses is relevant, because it is the main reason for climate-related sea-level rise, because glaciers are significant freshwater sources in Asia and because it is another negative feedback effect due to the reduced reflection of sunlight back into space (albedo effect) which reduces warming. Despite this new evidence, emissions are still increasing and a political commitment to mitigate climate change remains out of sight. It is worth asking who the important parties are and who is responsible for the fight against climate change. Of course, households and firms are important actors, because their economic activities are the ultimate cause of greenhouse gas emissions. Governments that provide the legal framework of economic activities are economic actors themselves and therefore highly relevant, too. In this book, we ask about the role and responsibility of science and especially those of economists.

## 1.2    Climate Change as an Economic Problem

By now, climate change is a problem that is at least as important for economics and other social sciences as it is for the natural sciences. Climate science has established the existence of anthropogenic climate change and understands many of its physical and chemical causes. We also have a rough idea how climate change might affect ecosystems. Yet climate change is not only a huge problem for nature and the ecosystems but also for the human society and a threat to human life. From a practical perspective, climate change is more a societal problem than a scientific one. Scientifically, one might argue that the solution of how to halt global warming is quite trivial. Since the accumulation of greenhouse gases in the atmosphere is the cause of global warming, we can solve the problem by stopping the emission of those gases. However, we do not know how our societies can achieve this without crashing our social systems. Similarly, while we do have some ideas about the physical consequences of a changing climate, it is hard to assess what the economic and societal effects will be. We can be sure that there will be economic costs and welfare losses, but the size and distribution of those welfare effects are open questions.

In this book, we argue that economic science as a whole does not do enough to understand the economic causes and consequences of climate change and to make a contribution to the societal efforts to mitigate the effects of climate change. Climate change is both a societal and a scientific challenge. In our view, science and economists have several responsibilities to deal with this issue. One responsibility consists in doing research on climate change from multiple perspectives, because the complex network of causes and effects cannot be understood adequately from a single research approach. Another responsibility is to provide climate policy recommendations that have practical relevance and take the difficulties related to the implementation of effective policies into account. We derive these responsibilities from ethical considerations related to the goals of science and explain in what respect and why economists in general currently do not fulfil their responsibility. When talking about economics, we refer to the group of academic economists. Contrary to practical economists working in private companies or for public authorities, they are not expected to fight for vested interests. Since freedom entails responsibility, academic economists are the appropriate audience for our reflections.

Climate change is different from most other societal challenges that are analysed in the different academic disciplines. Its first important characteristic is the likely severity of its impact on all life on Earth. Natural scientists agree that strong increases in the average global temperature will be life-threatening for many species, including humans (IPCC 2014). To date, we may not know exactly from which temperature the world will become inhabitable for the current species, but we know that there exist critical temperature regions. There are other threats to (human) life on earth, such as nuclear warfare and accidents, pandemics or terrorism, but climate change is special because of its complexity, which is the second important characteristic.

Climate change is a complex problem, both for society and for science. It is global in scope, affects present and future generations and can only be solved by global measures. It is pervasive in the sense that it affects many natural and societal systems, which interact in non-linear ways with each other. These interactions make it extremely hard to predict and control the effects of climate change. Climate change is strongly interwoven with other problems as one of their causes, but also as an effect. Changing climatic conditions are related to how land can be used. They affect food production, the availability of fresh water and human health. Deteriorating physical living conditions in some regions cause migration and potentially local and regional conflicts. Migration and conflicts, in turn, have economic and societal implications for the countries that are directly involved, but also for other countries that are indirectly affected, for example, as transit countries or trade partners.

Climate change should be a central topic in economics, because it is an undesired side effect of economic activities, such as production, trade and consumption. The growth of these activities, which is a main topic of economic inquiry and a key goal of economic policy, will aggravate climate change under the current conditions. Economic activities are affected by the changing climate, which will damage physical and natural capital as inputs to production. Resources needed to deal with climate-related damages are not available for other uses implying a welfare loss. The effects on societies' resources, capital and production will affect financial markets and will also have distributional effects, both within countries and across countries. No matter whether societies actively or passively adapt to climate change or make efforts to mitigate it, there will be economic consequences and questions of trade-offs, which most economists consider as their main expertise. Dealing with these trade-offs concerning how societal resources should be used and how costs and benefits related to climate change are distributed are difficult issues. Likewise, it is difficult to determine how countries can transform their economies and societies, either to mitigate climate change or to adapt to it. Such transformations take time due to inherited habits, infrastructures and institutions. They will also cause resistance, because of status quo bias, fear of the unknown and expected welfare losses of some stakeholders.

It is therefore not enough to *determine* policies that bring about the required changes. Society as a whole has to find ways to *implement* the transformative policies so that they become effective. Economists can make significant contributions to the urgently needed transformation toward a low-carbon economy. However, the discipline currently does not make the best use of its expertise and influence, as we will show in this book.

## 1.3   Shortcomings of Mainstream Climate Change Economics

We take the stand that the discipline of economics in general does not contribute enough to the understanding of climate change and the necessary economic and societal transformations. More specifically, we claim, first, that economists do not do enough research on climate change, and, second, that most of the influential existing research is of limited relevance.

It is important to say clearly that those criticisms refer to what we call the *neoclassical mainstream of economics*. In our perception, which is shared by others (e.g. Fourcade et al. 2015), economics is dominated by one school of thought with specific ontological, epistemological and methodological premises that determine what questions are asked and how they are answered. Rather than saying that the neoclassical way of doing economics is wrong, we argue that it is a problem of doing it predominantly in one way. Climate change is a complicated, multi-faceted topic. It is one goal of this book to substantiate the claim that it is too simplistic to think about climate change from the perspective of a single school of thought and neglecting other approaches. In particular, deriving policy recommendations just from one paradigm can be misleading and, in the case of climate change, even dangerous.

We are fully aware of the fact that there is a broad literature off the economic mainstream and, of course, that there is the field of ecological economics with its nuanced view on climate change. In this book, we focus on the economic mainstream because of its strong and dominant role in research, teaching and policy advice. Ecological economics—at least in its origins—does not belong to this mainstream (Anderson and M'Gonigle 2012). The mainstream is defined by a set of journals that are considered important, the core in teaching that is considered as "standard" and hence reflected in the most frequently used textbooks, and those economists who are influential policy advisors and opinion leaders. In contrast to other disciplines in the social sciences, there is a strong belief in this mainstream, that there is only one proper way of doing economics, which leaves little room for pluralism. It is important to notice that we do not argue for a replacement of mainstream economics with a single heterodox perspective but for more pluralism in economics and esteem for the variety of economic approaches.

## 1.4    NEOCLASSICAL POLICY ADVICE

Analysing real-world problems just from a single perspective can lead to wrong policy conclusions, if the problem has many different dimensions as it is the case with climate change. While this theoretically applies to all disciplines, this is especially relevant for economics. Neoclassical mainstream economics is a strong and dominating school of thought, which almost exclusively focuses on *efficiency*. The respective literature is full of papers that derive optimal policies whose objective is to generate (Pareto-) efficient outcomes. Mainstream climate economics is part of this tradition. We have a close look at the research of William D. Nordhaus, who was awarded the Sveriges Riksbank Price in Economic Sciences for 2018 "for integrating climate change into long-run macroeconomic analysis" (Royal Swedish Academy of Sciences 2018, p. 45). Being awarded the Nobel Prize, Nordhaus is probably the most influential climate economist. We treat him as a representative of the *integrated assessment modelling* (IAM) approach, by which scientific models of the biosphere and atmosphere are linked with economic models in a common modelling framework. The economic part of those integrated assessment models typically is a version of the Ramsey growth model, which is deeply rooted in neoclassical methodology. A key element of economic analysis of climate change in the spirit of Nordhaus and his integrated assessment modelling is that both the consequences of climate change and their mitigation are costly to society. A rational policy of dealing with climate change should hence weigh the benefits of mitigation (avoided consequences) against the costs of mitigation. An optimal policy is one that efficiently trades off these costs and benefits such that an intertemporal social welfare function is maximised.

The neoclassical search for efficient climate change mitigation is not just an academic exercise, but found its way into the realm of public policy. In July 2019, the German council of economic advisors published a special report titled *Setting out for a new climate policy* (SVR 2019). The key message of this report is that climate policy should rely on market-based instruments, such as a carbon tax or emission certificates, and avoid regulatory law with detailed emission targets. The executive summary of the report states:

> Any climate policy that ignores economic considerations is ultimately doomed to failure. ... The guiding principle here is that greenhouse gas emissions can be reduced economically efficiently if the next unit is saved

wherever this is the most cost-effective—irrespective of at what location, with which technology, in which industrial sector and by which polluter this is achieved. ... A coordination strategy guided by market-principles thus plays a key role in achieving the goal of a cost-effective transformation. ... On the other hand, detailed targets—especially those set for individual sectors within economies—stand in the way of effective solutions. Moreover, it is questionable whether they are fundamentally suited to achieving the general climate objectives.

The report frequently cites Nordhaus' work, in particular the conviction that there should be a cost-benefit analysis and rational risk management (SVR 2019, p. 22). The council of economic advisors also refers to Nordhaus' estimates of marginal costs of additional $CO_2$ emissions, the so-called *social cost of carbon*. In the neoclassical theory, the optimal policy would be to impose a price or tax on carbon emissions that equal the social cost of carbon. Similar to the German council of economic advisors, its French counterpart, the conseil d'analyse économique (CAE), published a paper in March 2019, that also favours a carbon tax as the best instrument to reduce emissions (Bureau et al. 2019).

## 1.5    Value Judgments

Efficiency as the guiding principle of climate policy is a normative criterion that should be clearly stated as such and that can be questioned. An alternative principle might be the precautionary principle of risk minimisation. As all experts regularly emphasise (see IPCC 2014), climate science is fraught with uncertainties. What we call climate is the result of a complex system and so are the effects of climate on ecosystems and on the economy. Complex systems are characterised by multiple feedback effects, nonlinearities and the potential of sudden regime shifts, which make them extremely hard if not impossible to predict. At the same time, climate scientists are certain that there are so-called tipping points in the earth system, which means that the system's properties might change radically once certain temperature thresholds are passed (Lenton et al. 2008). The probability that such a tipping point is reached and cause irreversible change in this century is very low, but not zero (IPCC 2014). A prominent example of such a regime shift is the collapse of the Gulf Stream in the Atlantic Ocean, which would have significant effects on the climatic conditions in Europe.[3] However, it is neither possible to predict when regime shifts will occur nor what exactly will happen then. In part, this inability is due to so far insufficient data and theoretical modelling still in progress. Yet it is important to acknowledge that complex systems also have an element of inherent unpredictability that can never be eliminated by better data or modelling. Some fundamental uncertainty will always remain in the description and prediction of these systems. This implies the possibility that climate change causes catastrophic events, which could cause extremely high damages to nature and mankind, but also that we can never quantify the likelihood or the size of the damages before the events actually happen. This is why they can also be referred to as *known unknowns*. However, there might also be *unknown unknowns*, so called *black-swan* events, whose impacts are not even imaginable. The inability to quantify the likelihood and the magnitude of possible damages with acceptable precision is a fundamental problem for cost-benefit analysis and hence for the derivation of optimal policies which are based on these kinds of analyses. The precautionary principle would demand that we do as much as possible in order to prevent these catastrophic events from happening, instead of aiming at efficient or optimal solutions.

The uncertain possibility of catastrophic consequences of climate change causes problems not only for scientific analysis but also for the derivation of policy-relevant conclusions from research and the related public communication. The problem is that very serious consequences of climate change are possible, but there is no scientific way to prove this or to provide reliable estimates of the size and magnitude of the effects. In this sense, it is also a philosophy and ethics of science question of how science should deal with uncertainty and the limits of predictability and knowledge. The public image of scientists and also their self-perception is one of calm, rational people, who base their statements on facts and scientific knowledge. Talking about potential future events that cannot be supported with accepted scientific methods can appear speculative and unscientific. Scientists who are warning of catastrophes run the risk of being vilified as "alarmists", "gloom-mongers", "cassandras" or "doomsters and gloomsters", which might harm not only their scientific reputation but the reputation of science in general. Furthermore, dire descriptions

---

[3] https://www.theguardian.com/environment/2018/apr/13/avoid-at-all-costs-gulf-streams-record-weakening-prompts-warnings-global-warming; https://www.dw.com/en/collapse-of-gulf-stream-poses-threat-to-life-as-we-know-it/a-37092246.

of catastrophic outcomes are sometimes called "climate porn" in the news media,[4] which would distance the public from the problem instead of activating people to change their behaviour. Finally, there is the perception of a role conflict of scientists who not only analyse their object of study but also call for political action and thus provide normative judgment about what society should do about climate change. Von Storch and Krauss (2013) argue that climate science is caught in a "credibility trap", because scientists too often mingled with politicians and trespassed in the domain of political activism. In contrast, Nelson (2013) argues that scientists can and should not view themselves as detached, neutral observers of reality. On the contrary, she calls scientists and economists in particular to action in order to prevent severe climate change.

The uncertainty related to climate change creates a trade-off for both society and for economics. One the one hand, insufficiently mitigated climate change might result in disastrous outcomes with incredible suffering and welfare loss. The likelihood that this will happen, however, might be small. On the other hand, it is possible that relatively mild countermeasures are enough to prevent large harm and that future societies will be able to adapt to changed climatic conditions without major reductions in welfare. If future damages of climate change are very large, there are good reasons to make large and expensive efforts today in order to avoid huge future harm. Since we do not know how serious climate change could be and where the tipping points for disastrous results are, tough mitigation policy today might be inefficient in the sense that the costs for current societies are unnecessarily high. Counterfactually or ex post, it might have been possible to avert disaster at much lower costs.

Hence, we can frame the situation as a classical choice between a so-called type I error and a type II error. Science and society may have the null hypothesis that a climate apocalypse will not happen. A type I error—which is called false positive or false alarm—occurs if the null hypothesis is incorrectly rejected. A type II error (false negative) is the inverse: the null hypothesis is wrong, but we fail to reject it. In the case of climate change, a type I error could mean that there is too much investment in mitigation, because overly alarmist scientists convinced policymakers to take decisive action. A type II error might have the consequence that the earth system passes critical thresholds due to global warming resulting in large welfare

[4] https://www.theguardian.com/commentisfree/2006/aug/03/theproblemwithclimateporn.

losses, because mitigation efforts were not sufficient. Reasons for insufficient mitigation might be political and societal obstacles to transformative action, but also underestimation of the problem by economists.

Economists are used to seek small type I errors in their empirical work. High statistical significance, which means low probabilities of type I errors, is typically seen as something positive, despite long-standing criticism against this practice (Ziliak and McCloskey 2004). The rationale behind seeking low type I errors in the context of econometric model testing is that economists want to be confident that an asserted causal relation actually exists in order to reduce complexity and prevent statements like "in the social world, everything is related to everything". From an epistemic perspective, such conservatism in finding the "truth" is laudable. However, in practical contexts, there might be good reasons to put a larger weight on small type II errors, for example, in medical diagnosis, where the consequences of not detecting a serious illness might be more severe than the effects of a false alarm.

It is a value judgment how we assess the importance of the two errors. Economists who argue that climate policy first and above all should be efficient accept a larger probability of a type II error. They implicitly accept that the aspired optimal climate policy results in too little mitigation with catastrophic outcomes. Economists arguing that climate policy should be precautionary accept a larger probability of a type I error, meaning that some resources ex post were wasted because of too much mitigation. The positivistic view on science still dominant in mainstream economics holds that science in general and economics in particular should be value-free and focus on the objective analysis of economic issues. The example of climate economics demonstrates that this view is untenable. Already the choice of the research question is based on a value judgment. We can follow the neoclassical tradition and ask under which conditions social welfare is maximised, how optimal climate change would look like and how it could be achieved. Alternatively, we could ask what the worst outcomes of climate change on nature and society might be and how they could be prevented. Whereas one might accept the view that it is the responsibility of others like philosophers and politicians to evaluate the implications of research findings, economists cannot delegate normative judgements in their choice of the research questions and the research framework adopted. Freedom of research means that researchers have a choice and a responsibility for what they think about and how they do it. As we will show in this book, the choice of a school of thought inevitably implies value judgments that need to be clearly stated and discussed. Similarly, the perception of the role of scientists in society is a normative issue and related to responsibility of science as we will show.

## 1.6    TRANSFORMATIVE SCIENCE

Gibbons et al. (1994) distinguish two modes of knowledge production. Mode 1 is academic, discipline-based knowledge production or fundamental research. It is driven by curiosity, the desire to find truth and to answer questions posed by the scientific community. Societal relevance and applicability of this kind of knowledge are not important criteria for this kind of research. Mode 1 research aims at the production of true knowledge as defined by purely scientific standards. The scientific community is separated from other societal actors. In contrast, mode 2 of knowledge production aims at knowledge that is socially robust and applicable. It is transdisciplinary and the quality and relevance of its output are subject not only to academic standards, but also to criteria set by non-academic stakeholders. Academic scientists collaborate with non-academic researchers and other societal actors. This typology of research can be linked to the two goals of science, the epistemic goals of knowledge creation and the practical goal of applying knowledge to real life problems.

The distinction between mode 1 and mode 2 originally was meant to describe how actual knowledge production evolved in modern societies. After World War II, the general perception of the role of science was that of mode 1:

> the university was the place for both educating the elite and preparing the future through fundamental research, and this professorial elite was the only one to be in a position to decide what to do and to judge the quality and relevance of what was done. (Laredo 2007, p. 20)

It was a "republic of scholars" with high autonomy, which also comprised the allocation of funds by science organisations managed by peers (Laredo 2007). In the 1970s, the role of science was rethought. It was discovered that fundamental knowledge does not easily disseminate to the wider society by itself and that more active efforts to valorise the knowledge produced by universities were needed. This led to mode 2 research and the *third mission* of universities, which has social, entrepreneurial and innovative dimensions (Montesinos et al. 2008). While the third mission is often understood to refer to further activities in addition to traditional fundamental research and academic education, it also has an impact on how these missions are done, and hence leads to more mode 2 research.

A further step away from the idea that the role of science is to focus on fundamental or blue-skies research is the concept of science contributing to the solution of grand societal challenges or even the concept of a *transformative science* (Schneidewind et al. 2016). *Transformative science* means that science should not only observe and analyse processes in society, but rather should actively transform society by initiating and catalysing change processes. While the specific concept of transformative science originates from Germany (Schneidewind and Singer-Brodowski 2014), similar ideas were put forward in other countries before. The then president of the American Association of the Advancement of Science, Jane Lubchenko, spoke in 1997 of a "New Social Contract for Science" (Lubchenco 1998). Against the backdrop of environmental problems and climate change, the "[New] Contract would reflect the commitment of individuals and groups of scientists to focus their own efforts to be maximally helpful [to society]" (Lubchenco 1998, p. 495). In exchange for (more) funding provided by society,

> [t]he Contract should be predicated upon the assumptions that scientists will (i) address the most urgent needs of society, in proportion to their importance; (ii) communicate their knowledge and understanding widely in order to inform decisions of individuals and institutions; and (iii) exercise good judgment, wisdom, and humility. (Lubchenco 1998, p. 495)

The participants at the North American Meeting held in 1998 in advance of the World Conference on Science in 1999 expressed the need of a "New Contract between Science and Society" (see UNESCO 1998) with contents similar to Lubchenko's New Contract. The European Union's research and innovation programme Horizon 2020 is based on the idea of a "close partnership between science and society, with both sides working together towards common goals" and the expectation that "scientists must invest in society and Europe's ideals, pairing scientific excellence with social awareness and responsibility"[5] in order to find answers to key societal challenges. These ideas are not welcomed by

---

[5] https://ec.europa.eu/programmes/horizon2020/sites/horizon2020/files/FactSheet_Science_with_and_for_Society.pdf.

everyone, in particular not by representatives of traditional research funding organisations. The president of the German Research Foundation DFG, Peter Strohschneider, published a fervent critique of the concept of transformative science, arguing that it would lead to an undue importance and an overstrain of science and a depolitisation of politics (Strohschneider 2014).

## 1.7    Pluralism in Economics

There is hence a normative debate in science and science policy about the role of science, which is also highly relevant for economics and a key pillar to our argumentation. In our perception, economists currently do not participate enough in this debate. We argue in this book that it is the responsibility of the scientific community of economists and also of each individual academic economist to reflect on the question of the role of economics in society. Every economist should ponder on what he or she contributes to society, positively or negatively. The contribution depends both on the topic that is researched as well as on the methods that are used and that determine the results and the applicability to real world. Naturally, there are different opinions among different people on what is good and relevant research. The conditions in science should be such that there is a permanent exchange of ideas and arguments among scientists of different views and a struggle for the better argument. The conditions include an openness for other positions, a certain degree of familiarity with them and the basic readiness to accept the possibility that one's own argument might be inferior to those of other people and that every truth claim is contingent and preliminary.

It is not a new finding that current economics does not feature this openness. In 1992, 47 well-known economists, among them the four Nobel Laureates, Franco Modigliani, Paul Samuelson, Herbert Simon and Jan Tinbergen, published an appeal in the American Economic Review, stating:

> We the undersigned are concerned with the threat to economic science posed by intellectual monopoly. Economists today enforce a monopoly of method and core assumptions, often defended on no better ground that it constitutes the 'mainstream'. Economists will advocate free competition, but will not practice it in the marketplace of ideas. (Hodgson et al. 1992)

Since then, the situation has not improved. There are many publications that describe the current state of economics and argue that it is not pluralistic enough (e.g. Sent 2003; Schiffman 2004; Dequech 2007; Davis 2007; Fullbrook 2008; Dobusch and Kapeller 2014; Heise 2017).

We believe that pluralism in economics is necessary, as it is a contribution to a fruitful competition of ideas. Such a competition is a prerequisite for the production of new insights and the ongoing challenge of what is believed to be known and thus the progress of a discipline. The permanent awareness that most knowledge is fallible should be a main characteristic of modern science. While most economists endorse positivism at the practical level of everyday research, they do not do so at the deeper epistemological level. Mainstream economists even shun epistemological and ontological discussions with non-mainstream researchers. While there are plenty of methodological criticisms of mainstream economics, it is very hard to find mainstream economists replying to these attacks and defending what they are doing. In fact, the majority of mainstream economists are dismissive or even hostile to discussions of economic methodology (Lawson 1994; Boland 2019; Drakopoulos 2016) and often simply ignore their critics. We argue in this book that there is a responsibility for economists to change their mindset and to open up towards more pluralism. A plurality of economic approaches is necessary to deal with climate change, which we also show to be a responsibility of economists. Economics as a discipline has so much to offer and contribute in terms of expertise and methods. To systematically neglect non-mainstream thinking is to always use the same tool for all problems, while there is a rich toolkit of tools available that are probably better suited.

## REFERENCES

Anderson, Blake, and Michael M'Gonigle. 2012. Does Ecological Economics Have a Future? *Ecological Economics* 84: 37–48.

Bevis, Michael, Christopher Harig, Shfaqat A. Khan, Abel Brown, Frederik J. Simons, Michael Willis, Xavier Fettweis, Michiel R. van den Broeke, Finn Bo Madsen, Eric Kendrick, Dana J. Caccamise, Tonie van Dam, Per Knudsen, and Thomas Nylen. 2019. Accelerating Changes in Ice Mass Within Greenland, and the Ice Sheet's Sensitivity to Atmospheric Forcing. *Proceedings of the National Academy of Sciences of the United States of America* 116 (6): 1934–1939.

Boland, Lawrence. 2019. On Economic Methodology Literature from 1963 to Today. In *The Impact of Critical Rationalism: Expanding the Popperian Legacy through the Works of Ian C. Jarvie*, ed. Raphael Sassower and Nathaniel Laor, 19–29. Cham: Palgrave Macmillan.

Bureau, Dominique, Fanny Henriet, and Katheline Schubert. 2019. Proposal for the Climate: Taxing Carbon Not People. *Notes du conseil d'analyse économique* 50 (2): 1–12.

Cheng, Lijing, John Abraham, Zeke Hausfather, and Kevin E. Trenberth. 2019. How Fast Are the Oceans Warming? *Science (New York, N.Y.)* 363 (6423): 128–129.

Davis, John B. 2007. The Turn in Recent Economics and Return of Orthodoxy. *Cambridge Journal of Economics* 32 (3): 349–366.

Dequech, David. 2007. Neoclassical, Mainstream, Orthodox, and Heterodox Economics. *Journal of Post Keynesian Economics* 30 (2): 279–302.

Dobusch, Leonhard, and Jakob Kapeller. 2014. Heterodox United vs. Mainstream City?: Sketching a Framework for Interested Pluralism in Economics. *Journal of Economic Issues* 46 (4): 1035–1058.

Drakopoulos, Stavros A. 2016. Economic Crisis, Economic Methodology and the Scientific Ideal of Physics. *The Journal fo Philosophical Economics: Reflections on Economic and Social* x (1): 28–57.

Dunlap, Riley E., and Aaron M. McCright. 2011. Organized Climate Change Denial. In *The Oxford Handbook of Climate Change and Society*, ed. John S. Dryzek, Richard B. Norgaard, and David Schlosberg, 144–160. Oxford: Oxford Univ. Press.

Farquharson, Louise M., Vladimir E. Romanovsky, William L. Cable, Donald A. Walker, Steven V. Kokelj, and Dmitry Nicolsky. 2019. Climate Change Drives Widespread and Rapid Thermokarst Development in Very Cold Permafrost in the Canadian High Arctic. *Geophysical Research Letters* 46 (12): 6681–6689.

Fourcade, Marion, Etienne Ollion, and Yann Algan. 2015. The Superiority of Economists. *Journal of Economic Perspectives* 29 (1): 89–114.

Fullbrook, Edward, ed. 2008. *Pluralist Economics*. London, New York: Zed Books.

Gibbons, Michael, C. Limoges, H. Nowotny, S. Schwartzman, P. Scott, and M. Trow. 1994. *The New Production of Knowledge: The Dynamics of Science and Research in Contemporary Societies*. London: Sage Publications.

Heise, Arne. 2017. Defining Economic Pluralism: Ethical Norm or Scientific Imperative. *International Journal of Pluralism and Economics Education* 8 (1): 18–41.

Hodgson, G., U. Mäki, and D. McCloskey. 1992. A Plea for a Pluralistic and Rigorous Economics. *American Economic Review* 82 (2): 25.

IPCC, 2014. Climate Change 2014: Mitigation of Climate Change. Contribution of Working Group III to the Fifth Assessment Report of the Intergovernmental

Panel on Climate Change [Edenhofer, O., R. Pichs-Madruga, Y. Sokona, E. Farahani, S. Kadner, K. Seyboth, A. Adler, I. Baum, S. Brunner, P. Eickemeier, B. Kriemann, J. Savolainen, S. Schlömer, C. von Stechow, T. Zwickel and J.C. Minx (eds.)]. Cambridge University Press, Cambridge, United Kingdom and New York, NY, USA.

IPCC, 2018. Global warming of 1.5°C. An IPCC Special Report on the impacts of global warming of 1.5°C above pre-industrial levels and related global greenhouse gas emission pathways, in the context of strengthening the global response to the threat of climate change, sustainable development, and efforts to eradicate poverty [V. Masson-Delmotte, P. Zhai, H. O. Pörtner, D. Roberts, J. Skea, P.R. Shukla, A. Pirani, W. Moufouma-Okia, C. Péan, R. Pidcock, S. Connors, J. B. R. Matthews, Y. Chen, X. Zhou, M. I. Gomis, E. Lonnoy, T. Maycock, M. Tignor, T. Waterfield (eds.)].

Laredo, Philippe. 2007. Revisiting the Third Mission of Universities: Toward a Renewed Categorization of University Activities? *Higher Education Policy* 20 (4): 441–456.

Lawson, Tony. 1994. Why Are So Many Economists So Opposed to Methodology? *Journal of Economic Methodology* 1 (1): 105–134.

Lenton, Timothy M., Hermann Held, Elmar Kriegler, Jim W. Hall, Wolfgang Lucht, Stefan Rahmstorf, and Hans Joachim Schellnhuber. 2008. Tipping Elements in the Earth's Climate System. *Proceedings of the National Academy of Sciences of the United States of America* 105 (6): 1786–1793.

Lubchenco, J. 1998. Entering the Century of the Environment: A New Social Contract for Science. *Science (New York, N.Y.)* 279 (5350): 491–497.

Montesinos, Patricio, Jose Miguel Carot, Juan-Miguel Martinez, and Francisco Mora. 2008. Third Mission Ranking for World Class Universities: Beyond Teaching and Research. *Higher Education in Europe* 33 (2–3): 259–271.

Nelson, Julie A. 2013. Ethics and the Economist: What Climate Change Demands of Us. *Ecological Economics* 85: 145–154.

Oreskes, Naomi, and Erik M. Conway. 2010. *Merchants of Doubt: How a Handful of Scientists Obscured the Truth on Issues from Tobacco Smoke to Global Warming.* London: Bloomsbury.

Royal Swedish Academy of Sciences. 2018. *Economic growth, technological change, and climate change: Scientific background on the Sveriges Riksbank Prize in Economics Sciences in memory of Alfred Nobel 2018.*

Sachverständigenrat zur Begutachtung der gesamtwirtschaftlichen Lage (SVR). 2019. *Aufbruch zu einer neuen Klimapolitik.* Sondergutachten Juli 2019. Wiesbaden.

Schiffman, Daniel. 2004. Mainstream Economics, Heterodoxy and Academic Exclusion: A Review Essay. *European Journal of Political Economy* 20 (4): 1079–1095.

Schneidewind, Uwe, and Mandy Singer-Brodowski. 2014. *Transformative Wissenschaft: Klimawandel im deutschen Wissenschafts- und Hochschulsystem*. 2nd ed. Marburg: Metropolis Verlag.

Schneidewind, Uwe, Mandy Singer-Brodowski, Karoline Augenstein, and Franziska Stelzer. 2016. Pledge for a Transformative Science: A conceptual framework. Wuppertal Paper 2016(191). Wuppertal.

Schoolmeester, T., H.L. Gjerdi, J. Crump, B. Alfthan, J. Fabres, K. Johnson, L. Puikkonen, T. Kurvits, and E. Baker. 2019. *Global linkages—a graphic look at the changing Artic*. Nairobi and Arendal: UN Environment and GRID-Arendal. http://www.grida.no

Sent, Esther-Mirjam. 2003. Pleas for pluralism. In:Real World Economics: A Post-Autistic Economics Reader, ed. Edward Fullbrook, 177–184. Anthem Press.

von Storch, Hans, and Werner Krauss. 2013. *Die Klimafalle: Die gefährliche Nähe von Politik und Klimaforschung*. München: Hanser.

Strohschneider, Peter. 2014. Zur Politik der Transformativen Wissenschaft. In *Die Verfassung des Politischen*, ed. André Brodocz, Dietrich Herrmann, Rainer Schmidt, Daniel Schulz, and Julia Schulze Wessel, 175–192. Wiesbaden: Springer.

UNESCO. 1998. The role of science and technology in society and governance: Toward a new contract between science and society. Executive Summary of the Report of the North American Meeting held in advance of the World Conference on Science. http://www.unesco.org/science/wcs/meetings/eur_albert

Wilkerson, Jordan, Ronald Dobosy, David S. Sayres, Claire Healy, Edward Dumas, Bruce Baker, and James G. Anderson. 2019. Permafrost Nitrous Oxide Emissions Observed on a Landscape Scale Using the Airborne Eddy-Covariance Method. *Atmospheric Chemistry and Physics* 19 (7): 4257–4268.

Xu, Yangyang, Veerabhadran Ramanathan, and David G. Victor. 2018. Global Warming Will Happen Faster Than We Think. *Nature* 564 (7734): 30–32.

Zemp, M., M. Huss, E. Thibert, N. Eckert, R. McNabb, J. Huber, M. Barandun, H. Machguth, S.U. Nussbaumer, I. Gärtner-Roer, L. Thomson, F. Paul, F. Maussion, S. Kutuzov, and J.G. Cogley. 2019. Global Glacier Mass Changes and Their Contributions to Sea-Level Rise from 1961 to 2016. *Nature* 568 (7752): 382–386.

Ziliak, Stephen T., and Deirdre N. McCloskey. 2004. Size Matters: The Standard Error of Regressions in the American Economic Review. *The Journal of Socio-Economics* 33 (5): 527–546.

# CHAPTER 2

# Importance of Climate Change in Economics

**Abstract** This chapter provides pleas by economists who share the authors' view that the discipline of economics should contribute more to fight climate change. Although economists have a powerful toolkit and expertise to design policies for accelerated emission reduction, economics does not pay enough attention to climate change. Roos and Hoffart provide bibliographic evidence that especially mainstream economics devotes only little research to climate change. A keyword-search in the Web of Science database shows that only very few articles appear in so-called top economics journals. Most research is published in specialised field journals, some of which have low ratings in journal-rankings and low impact-factors. Analysing more than 15,000 PhD theses written at 152 North American universities reveal that climate change is of little research interest for young researchers.

**Keywords** Climate research • Journal presence • Bibliometric analysis • Relevance in research • Climate economics

## 2.1 Introduction

In this chapter, we give a summary of public pleas by economists who argue that their discipline should contribute more to the understanding of the economics of climate change. According to them, economics has

M. Roos, F. M. Hoffart, *Climate Economics*, Palgrave Studies in Sustainability, Environment and Macroeconomics, https://doi.org/10.1007/978-3-030-48423-1_2

powerful tools of analysis and can provide important insights how the problem could be addressed. However, economists do not pay enough attention to climate change. We agree with this view and argue that especially in mainstream economics not enough research is devoted to climate change. In order to support this view, we assess the actual importance of climate change within the science of economics referring to the following quantitative indicators: (1) number of journal articles on climate change published in economic journals, especially in the top 5 journals; (2) number and ranking of economic journals dedicated to those topics; (3) doctoral dissertations in North America. With these indicators, we intend to look both at the current state in economics (1–2), and at the possible future research (3).

Of course, the claim that there is *not enough* research on a topic is a subjective judgement. To support our impression that climate change does not receive the attention it deserves, we compare it with other topics to put the numbers into perspective. We do not think that these other issues are not relevant for society or not worth conducting research on. They are meant to be yardsticks of comparison for the relative importance of climate change among other topics economists are interested in.

We are aware that the numbers we present do not reveal any information about the quality and content of research nor differentiate between mainstream and heterodox research. We do not intend to propose what the *optimal* amount of research on climate change would be. Not knowing the optimal amount of research does not invalidate our argument that climate change issues are strongly underrepresented in economics. An analogy can be drawn to Amartya Sen (2006), who proposed in his famous theory of justice that an ideal theory of justice is not needed to criticise an unjust state that requires improvement.

## 2.2    Voices on the Contribution of Climate Economic

In the lecture William D. Nordhaus gave when he received the Nobel Prize in 2018, he stated that climate change is not only a topic of great importance for humanity but also an ultimate challenge for economics. He classified climate change economics as one "of the many fields in economics" and compared it with recognised achievements, such as the economic growth theory or the general-equilibrium theory (Nordhaus 2019,

p. 1991). For Nordhaus, there is no doubt that climate change is a central concern for economists. He even suggests the contribution of science as one of four steps for combating climate change and demands that "[s]cientists must continue intensive research on every aspect from science and ecology to economics and international relations. Those who understand the issue must speak up and debate contrarians who spread false and tendentious reasoning" (Nordhaus 2019, p. 2013).

Nicolas Stern talked about how to think about the economics of climate change, when he received the Leontief Prize in 2011. He (2011, p. 11) emphasised that the "economics of climate change is a subject which involves issues of extraordinary importance—essentially, for many people, existential." According to him, economics is equipped with a great toolkit and expertise to address climate change, which is not fully used. Stern's suggestion to "marshal the whole range of our subject" (Stern 2011, p. 3) and to go beyond a price of carbon implies that current economic research is not enough.

In 2019, the initiative *Economists for Future* (Econ4Future) was started by the *Rethinking Economics* network in Germany. On their platform, they state that "[e]conomists have failed to push politicians to action with the help of scientific facts. In many economic lectures there is still the idea of economic 'pragmatists' and ecological 'dreamers'." (Economists for Future—Rethinking Economics 2019; Netzwerk Plurale Ökonomik 2019). More than 2500 individuals and organisations from all over the world[1] signed their open letter from late 2019. The movement unites "for an economics that takes the climate science seriously" and claims that "[d]espite some exceptions, the overall contribution from economists has been nowhere near commensurate with the magnitude of the problem" (Economists for Future 2019).

Oswald and Stern (2019) ask "why has the engagement of economists been so week?" and refer to climate economics. "Economics must be at the heart of serious analysis of the issues and of the public policy necessary to tackle [climate change]" (Oswald and Stern 2019, p. 2). Furthermore, "academic economists have contributed disturbingly little to discussions about climate change and are failing the world and their own grandchildren." To support the lack of research on climate change, they look at articles on climate change published in nine high-ranked economic mainstream journals and show that there are only very few. The result makes

---

[1] Last updated on April 04, 2020, https://econ4future.org.

them worry that future generations may accuse economists that they "stood silently by and continued in a narrow way to ignore climate-change issues and to write journal articles on topics of less importance" (2019, p. 12).

For Nordhaus, Stern and many other economists, there is no doubt that climate change is a serious problem. It has an economic dimension and is thus a topic for economics. Although economists have the necessary tools and skills, many criticise the low contribution from economics. Admittedly, over the last decades, an increasing number of economists have been devoting attention on climate change (Hsiang and Kopp 2018, p. 4), following the example of Nordhaus. However, this does not necessarily imply a high relevance within the field in terms of publications and research activities. Thus, we provide evidence for a rather low actual importance of climate change.

## 2.3   ACTUAL IMPORTANCE OF CLIMATE CHANGE IN ECONOMICS

### 2.3.1   *Journal Articles on Climate Change*

We searched the Web of Science (WOS) database for the terms *climate change*[2] and *global warming* that appeared either in the title or as a topic of an economic journal article between January 1957 and September 2019.[3] We concentrated on journal articles and excluded all other sources, such as book reviews or proceedings papers, because journal articles are currently seen as the most relevant output in economics. Similar to the calculation of Oswald and Stern (2019), we filtered for different keywords, disciplines and economic journals and went through the results by hand. There might be aspects of our bibliometric search that are debatable, which however do not change the main message of our results.

One of the first economic articles that appeared in the search was published in 1987 in the *American Economic Review* by two US economists (Kokoski and Smith 1987). In 1989, further articles were published (Adams 1989; Crosson 1989; Oliveira and Skea 1989). They appeared

---

[2] "Climate-change" and climate change.

[3] A search in topics might be too broad, as the term may appear in the abstract or keywords, although the article is mainly on other issues. Conversely, only looking at the titles we might neglect relevant papers.

more than 50 years after other disciplines published on climate change (e.g. Callendar 1938), which implies that economics started later than other disciplines to study climate change. Over the time, the terms *climate change* and *global warming* appeared 1277 and 157 times in the title; 5508 and 663 times in the topic of all economic journals listed in the WOB (Table 2.1). Over all disciplines, there are a total of 11,891 journal articles with the keywords in the title. In comparison, economic articles make up 12.06%, which is a relatively high share. Only *environmental studies* (33.87%) and *environmental sciences* (27.17%) reach a higher share. Since these WOS-categories are quite broad, the result is not surprising. To complete the top categories, economics (rank 3) is followed by *meteorology/atmospheric science* (10.75%), *geography* (8.00%) and *political science* (4.96%).

The picture changes, when looking at the top 5 economic journals.[4] *Climate change* and *global warming* appear 26 times in the title and 32 times as a topic, which represents 1.81% and 0.52% of all economic articles. Excluding the *American Economic Review* from the list reduces the counts to 2 (titles) and 1 (topic). There is 1 article with global warming as a topic in *Econometrica*, and 1 article each in the *Journal of Political Economy* and in the *Review of Economic Studies*. Out of 8858 articles that were published in the *American Economic Review* in the period between January 1957 and September 2019, *climate change* or *global warming*

**Table 2.1** Economic articles on climate change, 01.1957–09.2019

| Journals | Climate change | | Global warming | | All top 5 articles |
|---|---|---|---|---|---|
| | Title | Topic | Title | Topic | |
| American Economic Review | 17 | 30 | 7 | 1 | 8858 |
| Econometrica | 0 | 0 | 0 | 1 | 3675 |
| Quarterly Journal of Economics | 0 | 0 | 0 | 0 | 3017 |
| Journal of Political Economy | 0 | 0 | 1 | 0 | 3332 |
| Review of Economic Studies | 0 | 0 | 1 | 0 | 2823 |
| Σ top 5 articles | 17 | 30 | 9 | 2 | 18,882 |
| Σ all | 1277 | 5508 | 157 | 5508 | n.a |
| Share top 5 in all top 5 (in %) | 0.09 | 0.16 | 0.05 | 0.01 | |

Source: Authors' own contribution

[4] American Economic Review (AER), Quarterly Journal of Economics (QJE), Journal of Political Economy (JPE), Review of Economic Studies (RES) and Econometrica.

appears 24 times (0.27%) in the title and 31 times as a topic (0.35%). We interpret these shares as very small.

These results are in line with the findings of Oswald and Stern (2019), who searched in the top 9 economic journals for articles with the composite search term *Climate OR Carbon OR warming*. Compared to Oswald and Stern's WOS search from August 2019, our keywords are stricter. This difference explains why we got less search results, which implies a more pessimistic picture. In addition to the 5 top journals, they also looked at the *Journal of the European Economic Association, Economic Journal, Economica, American Economic Journal—Applied Economics.* They (2019, p. 10) conclude that "[a]cademic economics—at least as represented in mainstream general journals [...] has contributed remarkably few articles on one of the greatest scientific, economic and policy issues of our era". For them, the example of the *Quarterly Journal of Economics* is most shocking. It published more articles on basketball or baseball than on climate change (zero). The other journals' results are not very different.

In 2008, Goodall (2008) already identified a neglect of climate change research in the social sciences. She analysed 60 leading social science journals, which, according to her findings, failed to respond to climate change. A comparison between the top 30 journals of economics, political science and sociology revealed that the science of economics published the most articles. Goodall and Oswald criticised in the *Financial Times* (2019) that economic researchers "obsessed with FT Journals list are failing to tackle today's problems". They referred to the loss of biodiversity, which is related to climate change.

For a comparison, we searched in the top five journals for other, arguably less urgent terms such as *marriage/wedding* (title: 46, topic: 137), *schooling/education* (title: 266, topic: 453), *election/vote* (title: 134, topic: 361) and *corruption* (title: 19, topic: 49). As Table 2.2 shows, these keywords appear more often than *climate change* and *global warming*. It is important to note that we are not arguing that these topics are unimportant or not valuable for society. We rather think that there are topics that are more central to economics and most importantly more urgent, namely climate change. For reasons of completeness, we also searched for classical economic terms such as *Economic growth/GDP* (title: 190, topic: 290) or *taxation/tax* (title: 643, topic: 568), which also appear more often than *climate change* or *global warming*.

It is alarming that comparably little top research addresses the challenge of climate change. 98.19% of all economic articles with climate

**Table 2.2**  Economic articles on other topics than climate change, 01.1957–09.2019

| | Period of time: 01.01.1957–2019 | | | | | | | | All top 5 articles |
| | Keywords | | | | | | | | |
| | Marriage/wedding | | Schooling/education | | Economic growth/GDP | | Taxation/tax | | |
| Journals | Title | Topic | Title | Topic | Title | Topic | Title | Topic | |
| American Economic Review | 16 | 48 | 143 | 236 | 93 | 185 | 293 | 428 | 8858 |
| Econometrica | 3 | 17 | 15 | 0 | 19 | 0 | 41 | 80 | 3675 |
| Quarterly Journal of Economics | 5 | 19 | 24 | 96 | 43 | 105 | 89 | 141 | 3017 |
| Journal of Political Economy | 22 | 53 | 68 | 121 | 35 | 0 | 134 | 196 | 3332 |
| Review of Economic Studies | 3 | 16 | 16 | 0 | 30 | 74 | 86 | 151 | 2823 |
| Σ top 5 articles | 49 | 153 | 266 | 453 | 190 | 290 | 643 | 996 | 18,882 |
| Σ all | 585 | 2095 | 5058 | 18,446 | 6059 | 25,612 | 16,522 | 30,794 | |
| Share top 5 in all top 5 (in %) | 0.26 | 0.81 | 1.41 | 2.40 | 1.01 | 1.54 | 3.41 | 5.27 | |

Source: Authors' own contribution based on Web of Science (Clarivate Analytics)

change or global warming in the title are published in journals other than the top 5 journals. Most publications can be found in *Energy Policy, Ecological Economics, Environmental Resource Economics, Energy Economics and Journal of Environmental Economics and Management* (in descending order of publications).

### 2.3.2   Field Journals Devoted to Climate Change

Therefore, it is worth looking at the journal landscape and especially journals dedicated to climate change. According to the WOB database, there are[5] 363 economic journals[6] in 2019. In 2018, a total of 21,163 articles were published in these journals, which were cited 1,046,567 times. Interestingly, within 236 categories listed in the WOS in 2018, economics is the category with the most journals, followed by *mathematics* (313) and *biochemistry/molecular biology* (298). From all 19,247 journals, 1.89% are economic journals. Over the years, the number of economic journals more than doubled compared to 161 journals in 1997 (209 journals in 2008, 333 journals in 2012).

Concerning the number of published articles, economics is on rank 41 with 21,162 articles. Five times more articles are published in *material science* (114,027) (rank 1), followed by *chemistry* (71,724) and *engineering, electrical & electronic* (70,247). A similar picture emerges looking at the total citations. *Economics* is on rank 29 with 1,046,567 citations compared to *material science* (4,389,013) on rank 1, followed by *chemistry* (3,833,448) and *biochemistry & molecular biology* (3,759,966). This comparison reveals that although economics has the most journals, this does not apply for the number of articles and citations.

The landscape of economic journals is very differentiated and wide-ranging. While some journals cover different topics and schools of thoughts, many others journals are dedicated to specific topics. The total number of citations implies that these articles are cited by a relatively small group of economists and not by a broad audience. Following this logic, the existence of special journal dedicated to environmental issues can be assumed. Therefore, we searched in the WOB for economic

---

[5] Last update: 20. June 2019.

[6] Please note that on the website Scimago there is a total of 596 economic journals listed with the possibility to filter for WOS core journals https://www.scimagojr.com/. For our purpose, it is enough to consider the economic journals listed in the WOS.

**Table 2.3** Economic Journals dedicated to environmental issues

| Journal (key word: resource, energy, climate change, global warming, environment, ecological) | 1998 | | | 2008 | | | 2018 | | |
|---|---|---|---|---|---|---|---|---|---|
| | Total citations | Impact factor | Rank | Total citations | Impact factor | Rank | Total citations | Impact factor | Rank |
| 1  Annual Review of Resource Economics | n.a. | n.a. | n.a. | n.a. | n.a. | n.a. | 745 | 2.98 | 51 |
| 2  Australian Journal of Agricultural and Resource Economics | 19 | 0.63 | 57 | 266 | 0.72 | 111 | 1,036 | 1.37 | 162 |
| 3  Climate Change Economics | n.a. | n.a. | n.a. | n.a. | n.a. | n.a. | 366 | 0.64 | 296 |
| 4  Ecological Economics | 499 | 1.06 | 27 | 4,347 | 1.91 | 19 | 25,091 | 4.28 | 20 |
| 5  Economics of Energy & Environmental Policy | n.a. | n.a. | n.a. | n.a. | n.a. | n.a. | 227 | 2.03 | 98 |
| 6  Energy Economics | 143 | 0.28 | 122 | 1,459 | 2.25 | 15 | 15,850 | 4.15 | 24 |
| 7  Energy Journal | 285 | 0.64 | 56 | 943 | 1.73 | 28 | 2,880 | 2.46 | 64 |
| 8  Energy Policy | n.a. | n.a. | n.a. | n.a. | n.a. | n.a. | 47,238 | 4.88 | 13 |
| 9  Environment and Development Economics | n.a. | n.a. | n.a. | n.a. | n.a. | n.a. | 1,543 | 1.22 | 181 |
| 10  Environmental & Resource Economics | n.a. | n.a. | n.a. | 1,314 | 1.08 | 70 | 4,728 | 2.15 | 85 |
| 11  International Environmental Agreements-Politics Law and Economics | n.a. | n.a. | n.a. | n.a. | n.a. | n.a. | 757 | 2.31 | 73 |
| 12  Journal of Agricultural and Resource Economics | 96 | 0.39 | 103 | 378 | 0.41 | 167 | 1,030 | 1.19 | 184 |
| 13  Journal of Environmental Economics and Management | 1,152 | 1.47 | 15 | 2,564 | 1.73 | 27 | 6,354 | 4.18 | 23 |
| 14  Marine Resource Economics | n.a. | n.a. | n.a. | n.a. | n.a. | n.a. | 1,031 | 2.80 | 55 |
| 15  Resource and Energy Economics | 122 | 0.26 | 125 | 573 | 1.08 | 69 | 2,109 | 2.37 | 71 |
| 16  Review of Environmental Economics and Policy | n.a. | n.a. | n.a. | n.a. | n.a. | n.a. | 1,136 | 6.65 | 3 |
| 17  Water Resources and Economics | n.a. | n.a. | n.a. | n.a. | n.a. | n.a. | 191 | 1.81 | 115 |
| **Total number of journals** | **160** | | | **208** | | | **363** | | |
| **Share of key journals in %** | **4.38** | | | **3.85** | | | **4.68** | | |

Source: Authors' own contribution based on Web of Science (Clarivate Analytics)

journals with the following key words: *climate change, global warming, ecological, environment, resource, energy* (Table 2.3). Looking at three exemplary years, the resulting number of journals increased from 7 in 1998 to 8 in 2008 and 16 in 2018. Since the number of economic journals increased simultaneously, the share of journals dedicates to environmental issues did not change a lot (1998: 4.38%; 2008: 3.85%; 2018: 4.68%).

There might be other economic journals that fall within this category but do not fit the search criteria. Furthermore, articles on environmental issues might also be published in other journals which are not specifically dedicated to these issues. Still, it is worth having a closer look at the search results, as the main message remains the same. In 2018, the journal *Review of Environmental Economics and Policy* that is listed in the WOS since 2009 has the highest rank (3) based on the impact factor (6.65) among the key journals. It is also the only one among the top 10 economic journals in 2018 (see "Rank" in Table 2.3). Rank 1 is taken by the *Quarterly Journal of Economics* with an impact factor of 11.78. Five journals are, however, within the top 10% of the economic journals in 2018. The *Review of Environmental Economics and Policy* is one of these and has few citations (1136) compared to *Energy Policy* (47,238) on rank 13 or *Ecological Economics* (25,091) on rank 20. The latter two are the only ones ranked within the top 10% of economic journals in 2008, while none is within the top 10. Although impact factor and the citations of both journals increased from 2008 to 2018, their total rank decreased. Similarly, in 1998, there was no key journal among the top 10. The *Journal of Environmental Economics and Management* was ranked the highest (Rank 15). The journal *Climate Change Economics* exists since 2016 and is relatively new. Therefore, it is not surprising that its ranking (impact factor, total citations) is very low. It also reveals a need for a platform to publish articles on climate change, which is not met by the other journals.

### 2.3.3    PhD Theses on Environmental Aspects

Looking at the research of future economists enriches the picture. We think that doctoral studies reveal interesting information about what topics the next generation of economists finds interesting and career enhancing. It also shows how economic research might develop. Due to the following reasons, young researchers such as PhD students might be more interested in doing research on climate change compared to senior economists: First, they are not fully committed to a specific research topic or method yet, while senior economists hardly change their core research. Second, today climate change is more present in political discussions and the media compared to 20 years ago. When developing their research interest, they are hence more aware of climate change, which makes it an interesting topic of research for them. Finally, young researchers have a bigger personal interest in fighting climate change than older researchers, since they and their families will be more affected.

They will more likely be exposed to severe consequences of climate change.

Thus, we searched for different keywords related to climate change in the list of doctoral degrees awarded by US and Canadian universities, which is published for every academic year in the *Journal of Economic Literature*.[7] The North American context is interesting not only due to the availability of data, but also due to its leading role in the academic world. The Nobel Prize in economics is an interesting example. Out of 84 economics that were awarded with the Nobel Prize in economics since 1969, 68 were employed at US universities at the time they received the prize (Nobel Media AB 2019). Since the US has worldwide leading universities that dominate the academic economic discourse, it can be expected that some of their PhD students will step in these shoes and become influential economists and advisers themselves.

We combined data from 15 lists[8] that include a total of 15,312 dissertations from 152 North American universities from July 2001 until June 2018. Over all years, the keywords *climate change* (73) and *global warming* (4) appear in the titles of 77 dissertations which equal 0.50% of all dissertations. With the highest values in 2015 (16) and 2016 (10), the number of key doctoral studies is very low. These 77 doctoral degrees were granted by 42 different universities. It implies that at 72.37% of all universities that are represented in the list of PhD degrees, no PhD student focused on climate change.[9] Looking at the leading US universities for economics, 1 of the 5 top universities (University of Chicago: 1) and additional 4 of the top 10 Universities[10] (Princeton University: 1, Cornell University: 1, Yale University: 1, Columbia University: 2) are represented with only very few dissertations. Surprisingly, Yale University where Nobel laureate William D. Nordhaus is professor is not represented very often. The universities with the most doctoral studies on climate change are Texas

---

[7] The lists of degrees, subjects and recipients are provided by the universities.

[8] JEL (2002, 2003, 2004, 2005, 2006, 2007, 2008, 2009, 2010, 2012, 2013, 2014, 2015, 2016, 2017).

[9] According to the titles in the last 18 years.

[10] 1. Massachusetts Institute of Technology (MIT), 2. Stanford University, 3. Harvard University, 4. California Institute of Technology (Caltech), 5. University of Chicago, 6. Princeton University, 7. Cornell University, 8. Yale University, 9. Columbia University, 10. University of Pennsylvania, based on (15.10.2019).

A&M University (8), University of Alberta (4) and University of California, Berkeley (4).

Again, we are aware that there might be PhD theses that covered climate change, but do not appear in the search since the title does not contain the keywords. To get a more differentiated picture, we extended the search and created three categories containing the following keywords: *climate change, global warming (core category), emission, externalities, $CO_2$, greenhouse gas (extension I), sustainability, environment, energy (extension II)*. While the second category contains climate specific keywords, the third one comprises more general keywords. As displayed in Table 2.4, the keywords of the category *extension I* appear 111 times, which equals 0.72% of all dissertations, while *extension II* keywords appear in 565 titles (3.69%). In total, all 9 keywords appear in 753 titles, which make up 4.92% of all dissertations. Although the extension of keyword also increased the share from the core category of less than 1% to almost 5%, the share remains still very small.

As one could expect an increase on climate change research in the recent years, due to an increased urgency and media presence, we had a closer look at the distribution over time (Fig. 2.1).

The broader category 'extension II' still makes up the largest part. Taking all categories together, the appearance of keywords increased over time, starting with 13 search hits in 2001 and was the highest in the years 2011 (68), 2013 (72) and 2015 (85). This positive trend must be seen in

**Table 2.4**  Dissertations on climate change, North America (2001–2018)

| Category | Key word | Frequency | Share in total dissertations (%) |
| --- | --- | --- | --- |
| Core category | Climate change | 73 | 0.48 |
| | Global warming | 4 | 0.03 |
| Extension I | Emission | 35 | 0.23 |
| | Externalities | 63 | 0.41 |
| | CO2 | 4 | 0.03 |
| | Greenhouse gas | 9 | 0.06 |
| Extension II | Sustainability | 21 | 0.14 |
| | Environment | 397 | 2.59 |
| | Energy | 147 | 0.96 |
| SUM | | 753 | 4.92 |
| Total Dissertation | | 15,312 | 100 |

Source: Authors' own contribution based on JEL (2002–2018)

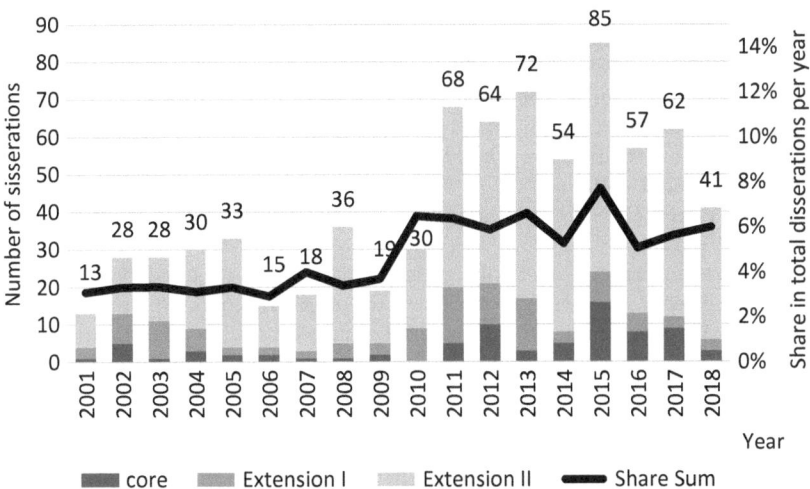

**Fig. 2.1** Dissertations on climate change over time (2001–2018). (Source: Authors' own contribution based on JEL [2002–2018])

relation to the number of dissertations per year, which also increased over time, but varies between 463 (2010) and 1124 (2016). Looking at the different years, the share rose from 3.10% in 2001, peaked in 2015 (7.72%) and was the lowest in 2006 (2.94%). Although the share in recent years is higher, the percentage of PhD students working on climate change is still very low.

In summary, our bibliometric search revealed that climate change is not a very prominent topic in economics, considering the urgency of the topic. This is especially true for articles published in the leading economic journals. The journal landscape reveals a similar picture. Comparatively few field journals are devoted to climate change. Most of those journals have no top ranking, which has implication on the articles' perceived quality. We find it remarkable that PhD students too, who represent the future economists, seem to follow this trend and concentrate only rarely on climate change in their PhD theses. Hence, our claim that there is not enough economic research devoted to climate change is supported both by voices from other economists and by bibliometric evidence. In Chap. 5, we will reflect about possible explanations for the lack of research on climate change.

# REFERENCES

Adams, Richard. 1989. Global Climate Change and Agriculture: An Economic Perspective. *American Journal of Agricultural Economics* 71 (5): 1272.

Callendar, Gyu Stuart. 1938. The Artificial Production of Carbon Dioxide and Its Influence on Temperature. *Quarterly Journal of the Royal Meteorological Society* 64 (275): 223–240.

Crosson, Pierre. 1989. Greenhouse Warming and Climate Change. *Food Policy* 14 (2): 107–118.

Economists for Future—Rethinking Economics. 2019. Accessed November 4, 2019. http://www.rethinkeconomics.org/projects/economists-for-future/.

Goodall, Amanda. 2008. Why Have the Leading Journals in Management (and Other Social Sciences) Failed to Respond to Climate Change? *Journal of Management Inquiry* 17 (4): 408–420.

Goodall, Amanda H., and Andrew Oswald. 2019. Researchers Obsessed with FT Journals List Are Failing to Tackle Today's Problems. *Financial Times*, May 8.

Hsiang, Solomon, and Robert Kopp. 2018. An Economist's Guide to Climate Change Science. *Journal of Economic Perspectives* 32 (4): 3–32.

JEL. 2002. Doctoral Dissertations in Economics Ninety-ninth Annual List. *Journal of Economic Literature* 40 (4): 1450–1475.

———. 2003. Doctoral Dissertations in Economics Hundredth Annual List. *Journal of Economic Literature* 41 (4): 1461–1483.

———. 2004. Doctoral Dissertations in Economics Ninety-Ninth Annual List. *Journal of Economic Literature* 42 (4): 1294–1318.

———. 2005. Doctoral Dissertations in Economics One-Hundred-Second Annual List. *Journal of Economic Literature* 43 (4): 1190–1218.

———. 2006. Doctoral Dissertations in Economics: One-Hundred-Third Annual List. *Journal of Economic Literature* 44 (4): 1166–1191.

———. 2007. Doctoral Dissertations in Economics One-Hundred-Fourth Annual List. *Journal of Economic Literature* 45 (4): 1197–1223.

———. 2008. Doctoral Dissertations in Economics One-Hundred-Fifth Annual List. *Journal of Economic Literature* 46 (4): 1155–1182.

———. 2009. Doctoral Dissertations in Economics: One-Hundred-Sixth Annual List. *Journal of Economic Literature* 47 (4): 1271–1298.

———. 2010. Doctoral Dissertations in Economics One-Hundred-Seventh Annual List. *Journal of Economic Literature* 48 (4): 1156–1183.

———. 2012. Doctoral Dissertations in Economics One-Hundred-Ninth Annual List. *Journal of Economic Literature* 50 (4): 1281–1310.

———. 2013. Doctoral Dissertations in Economics One-Hundred-Tenth Annual List. *Journal of Economic Literature* 51 (4): 1326–1355.

———. 2014. Doctoral Dissertations in Economics One-Hundred-Eleventh Annual List. *Journal of Economic Literature* 52 (4): 1309–1336.

———. 2015. Doctoral Dissertations in Economics One-Hundred-Twelfth Annual List. *Journal of Economic Literature* 53 (4): 1186–1215.

———. 2016. Doctoral Dissertations in Economics One-Hundred-Twelfth Annual List. *Journal of Economic Literature* 54 (4): 1551–1580.

———. 2017. Doctoral Dissertations in Economics One-Hundred-Fourteenth Annual List. *Journal of Economic Literature* 55 (4): 1761–1792.

Netzwerk Plurale Ökonomik. 2019. Economists for Future. Accessed November 4, 2019. https://plurale-oekonomik.de/projekte/economists-for-future/.

Nobel Media AB. 2019. Mon. 2019. All Prizes in Economic Sciences. Retrieved November 4, 2019. https://www.nobelprize.org/prizes/lists/all-prizes-in-economic-sciences/.

Nordhaus, William D. 2019. Climate Change: The Ultimate Challenge for Economics. *American Economic Review* 109 (6): 1991–2014.

Oliveira, Adilson, and Jim Skea. 1989. Global Warming—Time for a Cool Look. *Energy Policy* 17 (6): 543–546.

Oswald, Andrew, and Nicholas Stern. 2019. *Why does the Economics of Climate Change Matter so much, and Why has the Engagement of Economists been so Weak?* https://www.res.org.uk/resources-page/october-2019-newsletter-why-does-the-economics-of-climate-change-matter-so-much-and-why-has-the-engagement-ofeconomists-been-so-weak.html.

Sen, Amartya. 2006. What Do We Want From a Theory of Justice? *Journal of Philosophy* 103 (5): 215–238.

Stern, Nicholas. 2011. How should we Think about the Economics of Climate Change. *Lecture for the Leontief Prize*. Medord. http://www.ase.tufts.edu/gdae/about_us/leontief/SternLecture.pdf.

# CHAPTER 3

# Mainstream Climate Economics

**Abstract** In this chapter, Roos and Hoffart provide a formal exposition of William D. Nordhaus' DICE model. The DICE model is the workhorse model in mainstream climate economics. It triggered a lot of research, is used as a canonical model in teaching and is also the basis for policy recommendation. Especially the social cost of carbon, which can be calculated with models of the DICE type, plays an important role in climate policy, because the social cost of carbon is an estimation of the optimal carbon price. The authors discuss some extensions of the DICE model in the current literature and explain in which sense the DICE model represents economic mainstream thinking.

**Keywords** Carbon pricing • Climate economics • Climate policy • DICE model • Social cost of carbon

## 3.1 Introduction

In this chapter, we present the DICE model developed by William D. Nordhaus who was awarded the Sveriges Riskbank Prize in Economic Sciences in Honour of Alfred Nobel in 2018. The Nobel Committee awarded the prize for Nordhaus' path-breaking work in "integrating

© The Author(s) 2021
M. Roos, F. M. Hoffart, *Climate Economics*, Palgrave Studies in Sustainability, Environment and Macroeconomics,
https://doi.org/10.1007/978-3-030-48423-1_3

climate change into long-run macroeconomic analysis".[1] DICE stands for Dynamic Integrated model of Climate and the Economy. The first version was published in Nordhaus (1992) and Nordhaus and his co-authors updated and refined it several times (Nordhaus 2018). The DICE model is a so-called Integrated Assessment Model (IAM) that links a standard economic growth model with a model from climate science in order to analyse the interaction between the economy and the climate in an integrated modelling framework.

Instead of providing a comprehensive survey of the literature, we focus on the DICE model, because it is a canonical model that had an enormous impact both in science and in policy. The first version of the model sparked hundreds of follow-up papers. It also served as a teaching tool to familiarise generations of students with the economics of climate change (Gillingham 2018). Furthermore, it is simple and transparent enough so that it can be communicated to policymakers. In its relative simplicity, which makes it attractive for research, teaching and policy applications, the model resembles the Hicks-Hansen IS-LM model that popularised Keynesian ideas of the macroeconomy and that still serves as a starting point for discussion in most introductory textbooks in macroeconomics. As the Nobel Committee writes in its scientific background document, the DICE model and its regionalised version (RICE model) are still the "workhorse models for climate economics all over the world" (Committee 2018, p. 27). We therefore feel safe to treat the DICE model as representative for most of the economic mainstream thinking about climate change.

We would like to emphasise that we acknowledge the merits of William D. Nordhaus. Although we will criticise the mainstream approach of the economics of climate change and hence also the DICE model, this is not meant to be a personal attack against Nordhaus. On the contrary, Nordhaus deserves respect for his pioneering work and making both economists and policymakers aware of the challenge of climate change. Already, in 1974, Nordhaus predicted the atmospheric concentration of carbon dioxide in 2030 quite precisely, given current-day information (Gillingham 2018). He cautioned against melting polar ice caps 20 years before the IPCC did so (Barrage 2019) and subsequently introduced the idea of carbon taxes. Today, carbon taxes are supported by many economists; they are on the political agenda in many countries or even implemented at national, subnational or supranational levels (Barrage 2019). In January 2019, US

[1] https://www.nobelprize.org/prizes/economic-sciences/2018/nordhaus/facts/.

economists published a statement[2] calling for a carbon tax as the "most cost-effective lever to reduce carbon emissions".

DICE has been used by the US government to estimate the *social cost of carbon* (SCC) for regulatory impact analysis (Greenstone et al. 2013). The social cost of carbon is a measure of the marginal external costs caused by climate change due to the emission of carbon dioxide. It is the quantification of the externality that economic agents cause by emitting $CO_2$ when they produce, transport and consume goods. Carbon taxes correspond to the idea of Pigou taxes (Pigou 1920) that are intended to internalise external costs into the decision process of individual agents. In theory, if they are set at the right level, the individual considers the costs inflicted on others completely in her private comparison of the costs and benefits of an intended action. By considering the total costs, that is, private costs and social costs, the externality disappears.

The work of William D. Nordhaus can only be understood against the backdrop of Pigouvian levies to eliminate the externality of polluting the atmosphere with greenhouse gases,[3] especially $CO_2$ (Barrage 2019). The idea that such a price on pollutants of the atmosphere is needed was already mentioned in Nordhaus and Tobin (1972), five years after Nordhaus had received his PhD. The main problem of this idea was, and still is, that nobody could tell what carbon price was necessary to eliminate the externality. Most of Nordhaus' career was devoted to finding the appropriate level of the price on carbon emissions, for which economic considerations and scientific facts had to be integrated. In Nordhaus (1977), he presented the first preliminary estimates of carbon taxes and Nordhaus (1980) contains the first ever published numerical estimates of *optimal* carbon taxes based on an integrated cost-benefit analysis. According to Barrage (2019, p. 885), the development of an integrated modelling framework that allows to determine optimal climate policy was a "fundamental innovation". The Nobel Committee (2018) repeatedly emphasises that the Nordhaus models allow for welfare analyses and the identification of optimal, that is, cost-minimizing climate policy. As Barrage (2019, p. 886) writes, the work of Nordhaus "remains foundational for modern climate change economics, and carbon prices are now adopted by governments around the world to redirect the global economy

---

[2] https://www.econstatement.org.

[3] For simplicity, we will talk about $CO_2$ emissions in the following as a representation of all greenhouse gases.

towards a more sustainable long-term growth path. Without a doubt, this work exemplifies the spirit of Nobel's vision". Not only are his own models widely used, they are also the starting point and an inspiration for extensions and different IAMs of other authors.

## 3.2    FORMAL EXPOSITION OF THE DICE MODEL

We present the DICE-2010 version as described in Nordhaus (2013a). DICE is an aggregate model of the global economy, while the RICE model is a regionally disaggregated version of it. Since the logic of the RICE model follows the logic of the DICE model, we focus on DICE, which is simpler and more fundamental. Nordhaus states that the DICE model is primarily designed as a policy optimisation model that can be used to compare the outcomes and welfare implications of different policies. The model is intentionally kept simple in order to carve out the main interaction channels between the economic system and the climate system. As Barrage (2019, p. 898) notes, Nordhaus had already 20 years of experience in modelling the interaction between energy, climate and the economy when he presented the first version of DICE in 1992. His modelling choices about the relevant aspects of the problem are hence made very deliberately.

The basic structure of DICE is depicted in Fig. 3.1. The main idea is that there is a circular flow from $CO_2$ emissions originating from economic activity over rising temperature and climate impacts on the economy to climate-change policies that feed back into the economic system. As the question mark in Fig. 3.1 indicates, it is still unclear after all these years of research, which policies will and should be implemented globally and how they would affect the economy. The DICE model was designed to inform policymakers about the consequences of different policies.

### 3.2.1    Economic Part: Social Welfare and Economic System

In the following, we explain the economic part of the DICE model in more detail by going through its central equations. In essence, the economic part of DICE is a widely used model from the theory of optimal economic growth based on the works of Ramsey (1928), Cass (1965) and Koopmans (1965). It is assumed that global social welfare can be represented by a social welfare function $W$ that aggregates the individual utility levels of the world population over time:

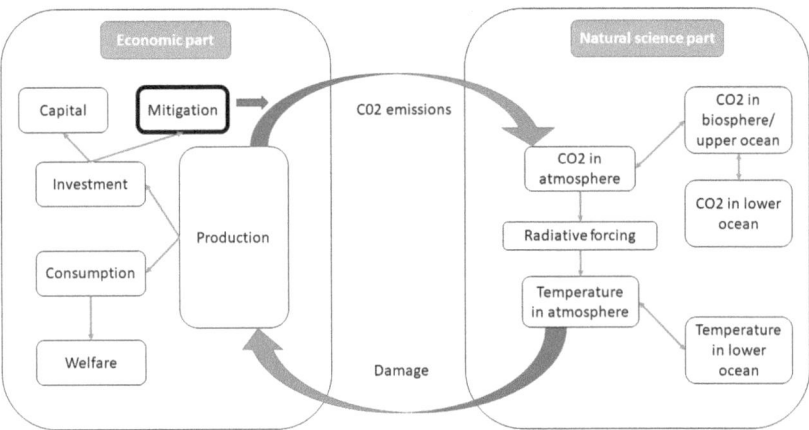

**Fig. 3.1** Basic structure of the DICE model

$$W = \sum_{t=1}^{T_{\max}} U\big[c(t),L(t)\big]R(t). \tag{3.1}$$

As standard in the macroeconomic growth literature, there is a representative consumer that obtains utility $U$ from the per capital consumption $c$. In DICE, *generalised consumption* is interpreted in a broad sense and includes not only goods and services traded in markets but also non-market items, such as leisure, health and environmental services. Nordhaus interprets the discrete time periods $t$ as *generations* such that the considered welfare is the weighted sum of the utilities of the generations 1 to $T_{\max}$. For the numerical computations, a specific functional form of the utility function is needed, which is assumed to be the widely used constant elasticity of substitution (CES) utility function:

$$U\big[c(t),L(t)\big] = L(t)\frac{c(t)^{1-\alpha}}{1-\alpha}. \tag{3.2}$$

The utility level of each generation depends only on this generation's per capital consumption. The individual utility levels are multiplied by the total population $L$ in this generation. The parameter $\alpha$ is called the *intertemporal elasticity of substitution*, which measures the substitutability of consumption of different generations.

To compute the social welfare of all generations considered, the generation utility levels are weighted with the *discount factor R*:

$$R(t) = \left(\frac{1}{1+\rho}\right)^t. \tag{3.3}$$

The parameter $\rho$ is called the pure rate of social time preference. If it is larger than zero, the utility levels of future generations are discounted in the welfare function, that is, they receive ever smaller weights approaching zero the further they are in the future.

The core idea of this approach is that the welfare function is maximised by the choice of relevant variables subject to the constraints given by the economic system and the climate system. Technically, the maximisation can be represented as a mathematical optimisation problem solved by a hypothetical social planner that chooses the control variables. In reality, of course, there is no global social planner that can do this. In line with the neoclassical literature, Nordhaus (2013a) argues that the market mechanism can be interpreted as a maximisation device and states that optimisation models can be seen as approximations of systems of interactive competitive markets. He refers to the report of an advisory committee (MRG 1978) chaired by Tjalling Koopmans, who is one of the pioneers of the optimal growth models and who received the Nobel Prize in 1975 for developing the theory of the optimum allocation of resources. Nordhaus emphasises that "we can interpret optimization models as a device for estimating the equilibrium of a market economy. ... [T]he maximization is an algorithm for finding the outcome of efficient competitive markets" (Nordhaus 2013a, p. 1111). The use of maximisation models hence presupposes that markets are competitive and the existence of an equilibrium state of the system.

The core of the economic system is the aggregate production function that describes where the goods to be consumed come from. Output $Q$ is produced with the input factors capital $K$, labor $L$ and energy, but energy only enters the function indirectly through the so-called *abatement function* $\Lambda$:

$$Q(t) = \left[1 - \Lambda(t)\right] A(t) K(t)^{\gamma} L(t)^{1-\gamma} \frac{1}{1 + \Omega(t)}. \tag{3.4}$$

The production is of the conventional Cobb-Douglas type and $A$ is total factor productivity. Since it is a growth model, $A$ and $L$ grow over time with exogenously given constant growth rates. The crucial innovation apart from the abatement function that Nordhaus introduced into the growth model is the *damage function* $\Omega$ that captures the damages caused by climate change expressed in lost output. The damage function comes in two versions:

$$\Omega(t) = f_1 \left[T_{AT}(t)\right] + f_2 \left[\text{SLR}(t)\right] + f_3 \left[M_{AT}(t)\right]. \tag{3.5}$$

$$\Omega(t) = \Psi_1 T_{AT}(t) + \Psi_2 T_{AT}(t)^2. \tag{3.5'}$$

Equation (3.5) is the complete damage function that is used in the RICE model. The damages due to climate change are caused by increasing atmospheric temperature $T_{AT}$, sea level rise SLR and $CO_2$ fertilisation due to atmospheric concentrations of carbon dioxide $M_{AT}$. The damages are of different kinds such as crop losses, damages to infrastructure due to weather events and sea level rise, adverse effects on health, non-market damages, as well as potential costs of catastrophic events. In the DICE model, the simplified version (3.5') is used in which the damage is a quadratic function of temperature only.

The abatement function $\Lambda$ measures the total costs of reductions of carbon emissions in terms of aggregate output:

$$\Lambda(t) = \pi(t) \theta_1(t) \mu(t)^{\theta_2}. \tag{3.6}$$

Hence, $1 - \Lambda$ is the fraction of total output left for consumption and investment after the costs of reducing emissions has been taken into account. The *emissions reduction rate* $\mu$ is a key control variable of the model. If the global society chose $\mu(t) = 0.5$, this would mean that technologies or behaviour changes are implemented that reduce carbon emissions by 50%. Nordhaus estimates the the polynomial function that describes the relation between abatement costs and the emissions reduction rate as highly convex, implying that the marginal reduction costs rise more than linearly with the reductions rate.

The economic model also has two standard macroeconomic accounting equations:

$$Q(t) = C(t) + I(t) \tag{3.7}$$

$$K(t) = (1 - \delta_K) K(t-1) + I(t). \tag{3.8}$$

Equation (3.7) says that, in market equilibrium, net output is used for total consumption $C(t) = c(t)L(t)$ and investment into capital goods $I$. Equation (3.8) is the capital accumulation equation stating that the capital stock of the current generation equals the capital stock of the previous generation net of depreciation plus the investment into new capital.

The key feature of DICE is the two-way interaction between the economy and the climate system. The effect of climate on the economy is captured by the damage function and the effect of the economy on the climate system results from the $CO_2$ emissions $E$. Total $CO_2$ emissions result from land use and from industrial emissions:

$$E(t) = E_{Land}(t) + E_{Ind}(t). \tag{3.9}$$

While land-use emissions $E_{Land}$ are assumed to be exogenous, industrial emissions $E_{Ind}$ depend on industrial production:

$$E_{Ind}(t) = \sigma(t) \left[ 1 - \mu(t) \right] A(t) K(t)^\gamma L(t)^{1-\gamma}. \tag{3.10}$$

If there is no reduction of the carbon emissions, that is, $\mu(t) = 0$, total emissions are the product of the exogenously given carbon intensity $\sigma$ and total ouput. The carbon emissions result from the use of fossil fuels, which is not directly modelled. However, the model contains a resource constraint that limits the sum of total carbon emissions to the total resources of fossil fuels $CCum$:

$$\sum_{t=1}^{T_{max}} E_{Ind}(t) \le CCum. \tag{3.11}$$

The model assumes intertemporally efficient resource use according to the Hotelling rule, with optimal resource rents and zero incremental extraction costs.

### 3.2.2    Natural Science Part: The Climate System

Nordhaus learned from physical and natural scientists at least since 1974 and chose a simplified but still adequate description of the complicated geophysical processes that determine the atmospheric temperature. His interdisciplinary work was an exception in economics at that time. In DICE, the climate system is represented by a model of the so-called carbon cycle (see Eqs. 3.12, 3.13, 3.14), an equation for the *radiative forcing* (see 3.15), and two equations for the temperature dynamics in the atmosphere and the lower oceans (see 3.16, 3.17).

The carbon cycle model describes how carbon is stored in the three reservoirs and how it flows between adjacent reservoirs. $M_{AT}$ is the amount of carbon that is stored in the atmosphere, which is the first reservoir. The second reservoir consists of the upper oceans and the biosphere and the amount of carbon stored there is $M_{UP}$. Finally, the deep oceans provide a vast sink for carbon in the long run, $M_{LO}$. Carbon emissions $E$ first flow into the atmosphere. From there, they can move first into the upper oceans and the biosphere, and later, rather slowly, into the lower oceans. The flows between adjacent reservoirs are two-directional. The three equations of the carbon cycle capture these dynamics:

$$M_{AT}(t) = E(t) + \phi_{11} M_{AT}(t-1) + \phi_{21} M_{UP}(t-1) \tag{3.12}$$

$$M_{UP}(t) = \phi_{12} M_{AT}(t-1) + \phi_{22} M_{UP}(t-1) + \phi_{32} M_{LO}(t-1) \tag{3.13}$$

$$M_{LO}(t) = \phi_{23} M_{UP}(t-1) + \phi_{33} M_{LO}(t-1). \tag{3.14}$$

The parameters $\phi_{ij}$ regulate the flows between the reservoirs and are calibrated to more complicated models and historical data.

The *radiative forcing* is a measure of the energy balance of the Earth due to the incoming sunlight. It is the difference between the sunlight absorbed by the Earth and the energy radiated back to space. The radiative forcing influences the temperature and depends on the concentration of greenhouse gases in the atmosphere. A simplified relation between the total change in radiative forcing since the pre-industrial base year 1750, $F$, and the accumulation of greenhouse gases is given by the following equation:

$$F(t) = \eta \left( \log_2 \frac{M_{AT}(t)}{M_{AT}(1750)} \right) + F_{EX}(t). \tag{3.15}$$

Since the most important anthropogenic greenhouse gas is $CO_2$, Eq. (3.15) explicitly considers the change of atmospheric $CO_2$ relative to the estimated pre-industrial level in 1750. The radiative forcing is also affected by other greenhouse gases such as methane, aerosols, ozone, and factors like albedo effects from clouds and glaciers. In DICE, all these forcing components are modelled as an exogenous factor $F_{EX}$, because they are either relatively small or poorly understood.

The final step that closes the model is to link the change in radiative forcing to temperature. If the radiative forcing increases, the temperature in the atmosphere $T_{AT}$ will rise. The warmer atmosphere warms up the upper ocean and, with a time lag, the temperature in the deep ocean $T_{LO}$ will rise, too, which is captured by the following dynamic equations:

$$T_{AT}(t) = T_{AT}(t-1) + \xi_1 \left\{ F(t) - \xi_2 T_{AT}(t-1) - \xi_3 \left[ T_{AT}(t-1) - T_{LO}(t-1) \right] \right\} \tag{3.16}$$

$$T_{LO}(t) = T_{LO}(t-1) + \xi_4 \left\{ T_{AT}(t-1) - T_{LO}(t-1) \right\}. \tag{3.17}$$

In a long-run equilibrium, the temperatures of the atmosphere and the deep ocean would not change. In that case, we find the long-run relation between the radiative forcing and atmospheric temperature to be $F = \xi_2 T_{AT}$ such that $\xi_2$ is the equilibrium temperature sensitivity.

## 3.3   Applications and Conclusions

Since 1992, Nordhaus updated the model and improved the calibration of the parameters to achieve a better fit to the historical data on output, $CO_2$ concentrations, temperature and the other endogenous variables. There is a host of results that have been produced with the different versions of the DICE/RICE model. For our purpose, it suffices to show some examples of the kind of outcome that DICE can produce. The computer code of the latest versions DICE-2013 and DICE-2016 and manuals on how the models can be used are available on Nordhaus' website,[4] which makes it possible for everyone to perform their own analyses with the models.

Figure 3.2 shows different policy cases climate economists like Nordhaus work with. It displays projections for the average global atmospheric temperature increase relative to the year 1900 for different cases. The four different cases correspond to different policy objectives and respective efficient policies to achieve these objectives.

Refering to the example of Nordhaus (2019), the first case represents the baseline case (Base), in which there are no policy interventions to mitigate climate change. It simply projects the future evolution of the system given the observed path of the emissions rate. Nordhaus (2019) projects that in the baseline case the average global temperature will rise steadily and will be 4 °C higher in 2100 than in 1900. This no policy case serves as a benchmark for the other cases.

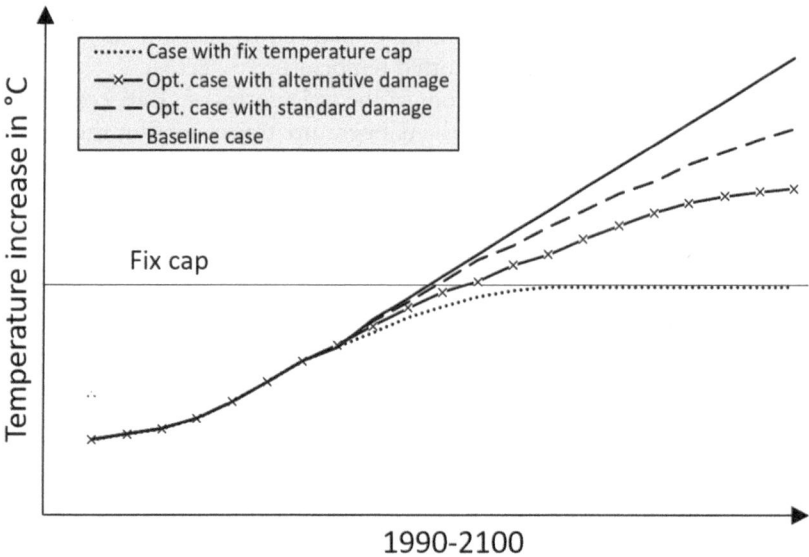

**Fig. 3.2**  Temperature trajectories in different policy cases. (Source: own figure)

[4] https://sites.google.com/site/williamdnordhaus/dice-rice.

There are two cases called *optimal*, describing the system with a cost-benefit optimum with the standard damage function (Opt) and an alternative damage function (alternative damage). In Nordhaus (2019), the damage coefficient of the alternative damage function is 3.5 times higher than in the standard case. In the optimal cases, the emissions rate is chosen such that the intertemporal welfare function (3.1) is maximised without any additional policy objective. As one can see in Fig. 3.2, an optimal policy with a standard damage function results in a much lower temperature increase than in the baseline case. In Nordhaus (2019), this case leads to a temperature increase of about 3.5 °C until 2100. If damages are more severe, however, the warming resulting from the optimal policy would be slightly below 3 °C in 2100. These two cases demonstrate that climate models such as DICE can be used to analyse the outcomes of different assumptions about the elements of the model. With a larger damage coefficient, global warming would cause higher damages and, hence, higher welfare losses. Consequently, the optimal policy would reduce the emissions rate more strongly, which would flatten the temperature trajectory.

In Nordhaus (2019), there is also a policy case corresponding to a hard cap with which the temperature is never allowed to pass a 2 °C increase. Figure 3.2 shows this case as a dotted temperature trajectory.

The DICE model can also produce the trajectories of GDP, consumption and the damages related to the temperature trajectories for the different policy cases. We focus here on the social cost of carbon, which would correspond to the optimal carbon tax in the different settings. Table 3.1 summarises the estimated social cost of carbon for the policy cases pathways in different years.

In the baseline case, in which no carbon tax equivalent to the SCC is implemented, the SCC rises from US$37 in 2015 to US$304 per ton of $CO_2$ in 2100. In the optimal case, the SCC is almost identical and serves as the optimal price on carbon implemented by policy measures. Note that the actual price on $CO_2$ in reality is far away from the optimal price in theory. First, we are far away from a global price of $CO_2$ that covers all sources of emissions. Second, even where a price is in force, it is too low. The $CO_2$ in the European system of emission allowances was about US$ 9/ton in 2015 and between US$ 21 and US$ 33 in 2019. As Table 3.1 shows, DICE estimates efficient carbon prices that are significantly higher, if global warming is to remain below 2 °C. With the hard cap, the prices are between US$ 225 in 2015 and US$ 459. If the 2°-target only holds

**Table 3.1** Estimations of the social costs of carbon based on the DICE 2016-R3 model

| | Social cost of carbon in 2018 $ per ton of $CO_2$ | | |
| --- | --- | --- | --- |
| | 2015 | 2020 | 2100 |
| Base | 37 | 45 | 303 |
| Optimal | 36 | 43 | 295 |
| Optimal (alternative damage) | 91 | 108 | 584 |
| $T \leq 2\ °C$ (100 year) | 130 | 158 | 1013 |
| $T \leq 2\ °C$ | 225 | 275 | 459 |

Source: adapted from Nordhaus (2019)

Notes: *Base* = no mitigation, *Optimal* = cost-benefit maximum, $T \leq 2\ °C$ *(x yr)* = temperature increase limited to 2 °C for a x-year average, $T \leq 2\ °C$ hard cap on 2 °C increase

over a period of 100 years, the price in 2015 could be lower at US$ 130, but would have to rise to US$ 1013 in 2100.

Such estimates of the SCC are the key result and the ultimate purpose of the DICE model and similar IAMs. William D. Nordhaus devoted his academic career to the estimation of optimal carbon prices based on the social costs that the externality of carbon dioxide emissions causes.

Figure 3.3 illustrates the basic reasoning of climate models, such as the DICE model. The rising solid line represents the damages cause by global warming, whereas the falling dotted line represents the mitigation costs. The U-shaped dashed curve is the sum of the two lines and represents the total of damages and mitigation costs. The socially optimal temperature increase $\Delta T^*$ is associated with the minimum of the total cost curve. Nordhaus (2013b), for example, estimates a minimum of the total costs of mitigated climate change of 2.9% of total income at 2.3 °C in a baseline case with efficient mitigation.

Different model variations are possible, one of which is shown in Fig. 3.4. It incorporates a tipping point in the climate system beyond which the damage function becomes very steep. The existence of a tipping point shifts the total cost curve to the left due to higher damage costs.

In this case, the cost-minimizing policy would limit the temperature increase just below the tipping point at a lower level than in the baseline case. At higher temperatures, the damages would be very high due to catastrophic events.

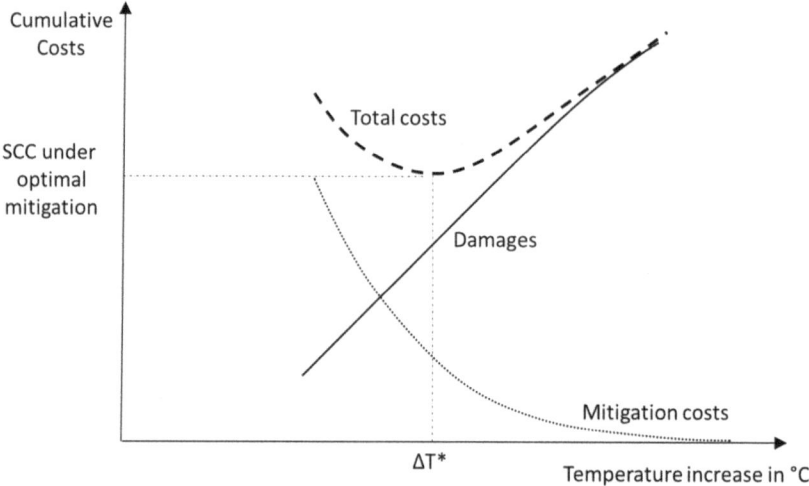

**Fig. 3.3**    Costs and benefits of mitigation policies. (Source: own figure)

Nordhaus (2013b) emphasises that we do not know whether Figs. 3.3 or 3.4 represents the more realistic case. There is huge uncertainty about the parameters of the climate and the economic system. We do not even know to a first approximation whether the assumed model parameters even have the right order of magnitude. This is why Nordhaus talks of a *climate casino*. Many extensions of DICE aim at understanding the consequences of this uncertainty and at exploring the implications of different assumptions concerning the highly uncertain model elements.

## 3.4    Extensions of DICE

One of the main issues that extensions of DICE deal with are different aspects of **uncertainty**, for example, the SLICE model of Kolstad (1993, 1994), with social learning or the probabilistic PRICE model (Nordhaus and Popp 1997).

Cai and Lontzek (2019) present the DSICE model, in which they jointly consider economics uncertainty about future productivity growth and climate uncertainty with regard to sensitivity parameters and the possible existence of tipping points. Several results of their analysis stand out. First, the SCC in the DSICE model is US$124/ton right at the start of

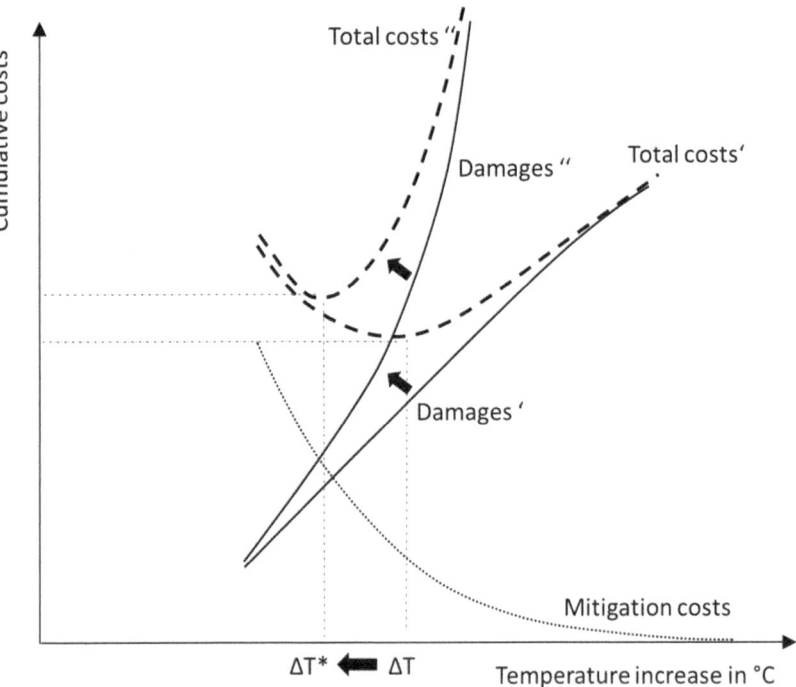

**Fig. 3.4** Costs and benefits of mitigation policies with a tipping point. (Source: own figure)

the simulation period in 2005, which is significantly higher than the SCC in the benchmark DICE with low damages. Second, the average SCC in 2100 is estimated to be US$ 461, which is even above the DICE SCC with the high damage coefficient. The most interesting observation, however, is the enormous variation of the estimates over the 10,000 simulation runs. In 2100, the SCC ranges from $100 (1% quantile) to $1700 (99% quantile). Cai and Lontzek (2019) find this large variation already for a single parameter set. They show that also considering uncertainty about parameters leads to even larger uncertainty about the SCC.

Nordhaus (2013a) distinguishes analyses of *second-order–moment uncertainty* (or thin-tailed uncertainty) and *higher-moment uncertainty*. Thin-tail uncertainty assumes that the tails of the distribution do not dominate the effects of uncertainty and that the distributions are normal or close to normal. His own analyses (e.g. Nordhaus 2008) rest on this assumption. His findings imply that the impact of this kind of uncertainty is relatively small such that "best-guess or certainty-equivalent policy seemed a good approximation for the policy in which a full expected-utility framework is used" (Nordhaus 2013a, p. 1117).

With higher-moment uncertainty, there is the potential of fat tails of the distribution and the risk of catastrophic tail events with extremely high impact. In the extreme, the expected loss from such catastrophic events could be infinite, for instance if global food production breaks down. Weitzman (2009) argues that we cannot perform standard cost-benefit analyses based on optimisation calculations with such tail risks, which he calls the *Dismal Theorem*. Nordhaus (2009) responds to Weitzman, that his derivation of the Dismal Theorem holds only under very specific assumptions which he considers irrelevant for a wide range of potential uncertain scenarios. However, he acknowledges that Weitzman raised an important point and calls for further theoretical and empirical research on the issue.

Another extension of DICE is to make **technological change endogenous**, which is done in the ENTICE model (Popp 2004), for example. Popp (2004) finds that ignoring induced innovation in the energy sectors as a consequence of climate change overstates the welfare costs of an optimal carbon tax policy by 9.4%. Endogenising technological change is relevant for policy analyses and the design of climate policies. Gillingham et al. (2008) provide an overview over the literature that models endogenous technological change for climate policy analysis.

Finally, the issue of **discounting** the future caused a major debate in the literature. In this well-known *Review on the Economic of Climate Change*, Nicholas Stern (2007) argued that the utility of future generations should be discounted very slightly or not even at all. Choosing low discount rates leads to much higher benefits of mitigation, because the future damages of climate change enter the social welfare function with larger weights in that case. Nordhaus (2007) vehemently defended the much criticised discounted rates chosen in his version of the DICE model. Subsequently, many other papers discussed the consequences of different choices for the discount rate and explored alternative modelling approaches (e.g. Schneider et al. 2012; Goulder and Williams 2012; Moxnes 2014; Arrow et al. 2013).

Summing up, the DICE model sparked a lot of debate and subsequent research in mainstream climate economics. A key finding of this literature overview is that alternative modelling approaches, especially concerning the treatment of uncertainty and the treatment of future utility, can have a huge impact on the estimated social cost of carbon and hence the optimal climate policies. We will elaborate more on this result in Chap. 4.

## 3.5   What Makes DICE Mainstream Thinking?

The DICE model is deeply rooted in mainstream economic thinking. This becomes clear when looking at Nordhaus' academic background, as well as the model's foundation, assumptions and purpose. Nordhaus earned his PhD at the MIT in 1967. Two of his PhD supervisors were later Nobel laureates: Robert Solow, who is one of the founders of neoclassical growth theory, and Paul Samuelson. Samuelson received the Nobel Prize in 1970 for his contribution to raising the analytical and methodological level in economic science. He formalised economics research using mathematics, and his work influences practically all branches of modern economics.[5] Apart from his many scientific articles, Samuelson's impact on rewriting economics also came from his famous textbook. The Nobel Committee wrote in 1970: "His *Economics: An Introductory Analysis*, first published in 1948, has become the best selling economics textbook of all time. ... Some economists feel that Samuelson's book ... is really his greatest contribution. It has gone a long way toward giving the world a common economic language".[6] When Samuelson received the Nobel Prize, his textbook was in its fifth edition. He wrote it as single author until the twelfth edition from 1985. It is important for our context that the newer editions were co-authored by William D. Nordhaus. On the Amazon website, the 19th edition is praised as follows: "The book continues to be the standard-bearer for principles courses, and this revision continues to be a clear, accurate, and interesting introduction to modern economics principles. Bill Nordhaus is now the primary author of this text, and he has revised the book to be as current and relevant as ever."

According to Arnsperger and Varoufakis (2013), neoclassical economics, which constitutes the current mainstream, rests on three axioms: (1) *methodological individualism*, (2) *methodological instrumentalism* and (3) *methodological equilibration*. The DICE model has all these features.

*Methodological individualism* means that there is a separation of structure from agency. All social structure is the result of individual behaviour

[5] https://www.nobelprize.org/prizes/economic-sciences/1970/samuelson/facts/.
[6] https://www.nobelprize.org/prizes/economic-sciences/1970/samuelson/facts/.

so that the individual agent is the natural starting point for every economic analysis. Ultimately, all explanations of socio-economic phenomena must be derived from the individual level. Since the 1970s, mainstream macroeconomists require that macroeconomic models have "rigorous microfoundations" and reject all other approaches that do not derive aggregate outcomes from individual behaviour (Roos 2017). The aggregation of individual decisions into a macro-level outcome is achieved by the fiction of the representative agent or consumer, which Nordhaus (1992) in line with most macroeconomists interprets as the "average individual".

*Methodological instrumentalism* implies that "all behaviour is preference driven or, more precisely, it is to be understood as a means for maximizing preference satisfaction" (Arnsperger and Varoufakis 2006, p. 8). In the DICE model, the welfare function (3.1) that is maximised represents aggregate individual preferences. While Nordhaus uses the social planner approach for simplicity, standard textbooks on economic growth theory (e.g. Barro and Sala-i-Martin 2004) or macroeconomic theory (e.g. Wickens 2012) show that the social planner approach is equivalent to the so-called decentralised approach. In this approach, there is a market equilibrium, with utility-maximising households and profit-maximising competitive firms. Nordhaus (2013a) endorses this interpretation.

Finally, *methodological equilibration* refers to the axiomatic imposition of equilibrium. In other words, whether or not a market equilibrium occurs is not the endogenous outcome of the models, but it is assumed that there is an equilibrium. Furthermore, neoclassical models do not only assume the existence of an equilibrium, they also assume that the equilibrium is unique and stable. Being in equilibrium or close to it is regarded as the natural state of the economy. In line with this thinking, Eq. (3.7) of the DICE model states that there is a market clearing equilibrium of produced output and aggregate demand. The implied Hotelling rule related to Eq. (3.11) relies also on an equilibrium assumption. As said before, the use of the social planner approach implicitly makes the assumption that the decentralised market economy is in equilibrium. All the trajectories of the endogenous variables that DICE generates, are succession of economic equilibria in the respective periods.

Apart from these conceptual features, DICE is a mainstream model in terms of the specific assumptions about the functional forms, for example, the production function or the welfare function. It is revealing that Nordhaus says about the welfare function: "This representation is a standard one in modern theories of optimal growth" (Nordhaus 2013a, p. 1081). With regard to output generation and capital accumulation, he states: "The economic sectors are standard to the economic growth literature" (Nordhaus 2013a, p. 1082). The aggregate Cobb-Douglas-type production functions with a variable called "capital stock" are distinctive marks of neoclassical growth theory (Foley et al. 2019).

At a more general level, DICE is representative for economic mainstream thinking with regard to its purpose. Right from the beginning of his career, it was Nordhaus' objective to build a model that can be used to determine optimal climate policies, which is also praised as his great contribution to the literature. Models that derive optimal economic policies abound in the mainstream literature. A search on Google Scholar with the keywords "optimal", "economic", and "policy" delivers about 3,450,000 hits[7] on topics such as "optimal monetary policy", "optimal trade and industrial policy", "optimal default", "optimal fiscal and growth policy" or "optimal taxation". Boldyrev and Ushakov (2016) state that general equilibrium modelling in the spirit of Kenneth Arrow, Gérard Debreu and Lionel McKenzie (ADM) had for a long time been seen as one of the "most significant intellectual constructions of neoclassical economics" (Boldyrev and Ushakov 2016, p. 39). Citing Mirowski (2002), they argue that the "ADM model … was expected to provide the most general framework for global planning and control that eventually would help to resolve the enormous computational problems associated with the operation of the economy and with economic policy" (Boldyrev and Ushakov 2016, p. 43). In this context, it is important to emphasise again the role of Tjalling Koopmans. He is one of the fathers of the Ramsey-Cass-Koopmans model, which is the economic core of DICE. Koopmans was engaged in planning problems at the so-called Cowles Commission and "saw the future of economics in engineering and programming (the very term revealing an affinity to computing problems, but also to planning and control) and saw theoretical economists as those who should provide the most fundamental principles of planning further elaborated by some applied analysts" (Boldyrev and Ushakov 2016, p. 43). They also mention the role of Samuelson, who formulated *the* economic problem as one of constrained optimisation. General equilibrium models are then natural optimisation problems that deal with the coordination of individual consumption and production plans. Neoclassical general equilibrium thinking

[7] 26 September 2019.

is hence more about how the world could be shaped by adequate mechanism design than about how it can be represented and explained. Although many contemporary mainstream economists might not be aware of the intellectual background on which they build their models, this attitude of social engineering expressed in the DICE model is pervasive in mainstream thinking.

## References

Arnsperger, Christian, and Yanis Varoufakis. 2006. What Is Neoclassical Economics? The Three Axioms Responsible for Its Theoretical Oeuvre, Practical Irrelevance and, This, Discursive Power. *Panoeconomicus* 1: 5–18.

———. 2013. Neoclassical Economics: Three Identifying Features. In *Pluralist Economics*, ed. Edward Fullbrook. London: Zed Books.

Arrow, Kenneth J., Maureen L. Cropper, Christian Gollier, Ben Groom, Geoffrey M. Heal, Richard G. Newell, William D. Nordhaus, Robert S. Pindyck, William A. Pizer, Paul R. Portney, Thomas Sterner, Richard S. J. Tol, and Martin L. Weitzman. 2013. *How Should Benefits and Costs Be Discounted in an Intergenerational Context? The Views of an Expert Panel* (December 19). Resources for the Future Discussion Paper No. 12-53. https://doi.org/10.2139/ssrn.2199511.

Barrage, Lint. 2019. The Nobel Memorial Prize for William D. *Nordhaus. Scandinavian Journal of Economics* 121 (3): 884–924.

Barro, Robert J., and Xavier Sala-i-Martin. 2004. *Economic Growth*. Cambridge: MIT Press.

Boldyrev, Ivan, and Alexey Ushakov. 2016. Adjusting the Model to Adjust the World: Constructive Mechanisms in Postwar General Equilibrium Theory. *Journal of Economic Methodology* 23 (1): 38–56.

Cai, Yongyang, and Thomas S. Lontzek. 2019. The Social Cost of Carbon with Economic and Climate Risks. *Journal of Political Economy* 127 (6): 2684–2734.

Cass, David. 1965. Optimum Growth in an Aggregative Model of Capital Accumulation. *Review of Economic Studies* 32 (3): 233–240.

Committee for the Prize in Economic Sciences in Memory of Alfred Nobel. 2018. *Economic Growth, Technological Change, and Climate Change.*

Foley, Duncan K., Thomas R. Michl, and Daniele Tavani. 2019. *Growth and disTribution*. Cambridge: Harvard University Press.

Gillingham, Kenneth. 2018. *William Nordhaus and the Costs of Climae Change*. VOX CEPR Policy Portal, October 19. https://voxeu.org/article/william-nordhaus-and-costs-climate-change.

Gillingham, Kenneth, Richard G. Newell, and William A. Pizer. 2008. Modeling Endogenous Technological Change for Climate Policy Analysis. *Energy Economics* 30 (6): 2734–2753.

Goulder, Lawrence H., and Roberton C. Williams III. 2012. The Choice of Discount Rate for Climate Change Policy Evaluation. *Climate Change Economics* 3 (4): 1250024.

Greenstone, M., E. Kopits, and A. Wolverton. 2013. Developing a Social Cost ofCarbon for US Regulatory Analysis: A Methodology and Interpretation. *Review of Environmental Economics and Policy* 7: 23–46.

Kolstad, Charles D. 1993. Looking vs. Leaping: The Timing of COs Control in the Face of Uncertainty and Learning. In *Costs, Impacts, and Benefits of CO 2 Mitigation, IIASA Collaborative Paper CP-93-002*, ed. Y. Kaya, N. Nakicenovic, W.D. Nordhaus, and F.L. Toth. Laxenburg: IIASA.

———. 1994. George Bush Versus Al Gore: Irreversibilities in Greenhouse Gas Accumulation and Emission Control Investment. *Energy Policy* 22: 771–778.

Koopmans, Tjalling. 1965. On the Concept of Optimal Economic Growth, in The Econometric Approach to Development Planning. *Pontif. Acad. Sc. Scripta Varia*. 28: 225–300. reissued North-Holland Publ. [1966].

Mirowski, Philip. 2002. *Machine Dreams: Economics Becomes Cyborg Science.* New York: Cambridge University Press.

Moxnes, Erling. 2014. Discounting, Climate and Sustainability. *Ecological Economics* 102: 158–166.

MRG. 1978. *Report of the Modeling Resource Group of the Committee on Nuclear and AlternativeEnergy Systems: Energy Modeling for an Uncertain Future.* Washington, DC: National Academy of Sciences Press.

Nordhaus, William D. 1974. Resources as a Constraint on Growth. *American Economic Review* 64 (2): 22–26.

———. 1977. Economic Growth andClimate: The Carbon Dioxide Problem. *American Economic Review* 67 (1): 341–346.

———. 1980. *Thinking About Carbon Dioxide: Theoretical and Empirical Aspects of Optimal Control Strategies.* Cowles Foundation Discussion Paper 565, Yale University.

———. 1992. An Optimal Transition Path for Slowing Climate Change. *Science* 20: 1315–1319.

———. 2007. A Review of the Stern Review on the Economics of Climate Change. *Journal of Economic Literature* 45 (3): 686–702.

———. 2008. *A Question of Balance: Weighing the Options on Global Warming Policies.* New Haven, CT: Yale University Press.

———. 2009. *An Analysis of the Dismal Theorem.* Cowles Foundation Discussion Paper No 1689, January 16.

———. 2013a. Integrated Economic and Climate Modeling. In *Handbook of Computable General Equilibrium Modeling SET*, ed. Peter B. Dixon and Dale W. Jorgenson, vol. 1B, 1069–1131. Amsterdam: Elsevier.

———. 2013b. *The Climate Casino—Risk, Uncertainty, and Economics for a Warming World*. New Haven and London: Yale University Press.

———. 2018. Evolution of Modeling of the Economics of Global Warming: Changes in the DICE Model, 1992–2017. *Climatic Change* 148: 623–640.

———. 2019. Climate Change: The Ultimate Challenge for Economists. *American Economic Review* 109 (6): 1991–2014.

Nordhaus, William D., and David Popp. 1997. What Is the Value of Scientific Knowledge? An Application to Global Warming Using the PRICE Model. *The Energy Journal* 18 (1): 1–45.

Nordhaus, William D., and James Tobin. 1972. Is Growth Obsolete? In *Economic Research: Retrospect and Prospect*, vol. 5, 1–80. Chicago, IL: Economic Growth, NBER Books.

Pigou, Arthur C. 1920. *Economics of Welfare*. London: Macmillan and Co.

Popp, David. 2004. ENTICE: Endogenous Technological Change in the DICE Model of Global Warming. *Journal fo Environmental Economics and Management* 48 (1): 742–768.

Ramsey, Frank P. 1928. A Mathematical Theory of Saving. *Economic Journal* 38: 543–559.

Roos, Michael. 2017. Behavioral and Complexity Macroeconomics. *European Journal of Economics and Economic Policies: Intervention* 14 (2): 18–199.

Schneider, Maik T., Christian P. Traeger, and Ralph Winkler. 2012. Trading Off Generations: Equity, Discounting, and Climate Change. *European Economic Review* 56 (8): 1621–1644.

Stern, Nicholas. 2007. *The Economics of Climate Change—The Stern Review*. Cambridge: Cambridge University Press.

Weitzman, Martin L. 2009. On Modeling and Interpreting the Economics of Catastrophic Climate Change. *Review of Economics and Statistics* 91: 1–19.

Wickens, Michael. 2012. *Macroeconomic Theory. A Dynamic General Equilibrium Approach*. Princeton, NJ and Oxford: Princeton University Press.

# CHAPTER 4

# What's Problematic About Mainstream Climate Economics?

**Abstract** Mainstream climate economics has a very narrow focus on the problem of climate change and a restricted set of potential policies for climate change mitigation. The neoclassical interpretation of climate change is that of an externality problem such that the appropriate policy is to impose a tax on carbon emissions in order to correct the market failure. Roos and Hoffart present ten alternative economic schools of thought, which have different perspectives on climate change and the economy in general. Those schools of thought emphasise other aspects of the climate change and analyse different questions. They hence also arrive at different conclusions about policy recommendations. The chapter presents a number of those questions which have not received sufficient attention in mainstream economics so far.

**Keywords** Critique neoclassical economics • Methodology • Epistemology • Normativity • Schools of thought

## 4.1 Introduction

In this chapter, we criticise mainstream climate economics in the tradition of the DICE model. Our critique is threefold.

© The Author(s) 2021
M. Roos, F. M. Hoffart, *Climate Economics*, Palgrave Studies in Sustainability, Environment and Macroeconomics,
https://doi.org/10.1007/978-3-030-48423-1_4

First, there should be more pluralism in economics. Although there is plurality of approaches in climate economics, there is a lack of *interested pluralism*. Due to the dominance of mainstream economics, the plurality of heterodox approaches is not visible and not valued as *proper* research. In Sect. 4.2, we discuss the meaning of interested pluralism and characterise a number of non-mainstream schools of thought. This presentation prepares the ground for the following critique of neoclassical climate economics.

Second, neoclassical climate economics does not adequately capture the complexities of the relation between climate change, the economy, society and public policy. It asks only very special kind of questions and provides answers that can be disputed because of the methodology used to produce them. Many issues important to society and for public policy are not addressed in neoclassical climate economics. In Sect. 4.3, we discuss the methodological weaknesses of the neoclassical approach to climate economics.

Finally, the dominance of the neoclassical approach has implication for public policy. It asks very specific questions and has a clear focus on market-based instruments to address climate change, such as $CO_2$ pricing. However, these instruments are not the only and not per se the best solution. Other aspects that are at least as important remain untouched and invisible for policymakers. Some neglected issues and their implications for climate policy are discussed in Sect. 4.4.

There has been too much model fine-tuning in climate economics over the past 30 years because of fundamental issues with neoclassical modelling. A lot of scientific effort went into the improvement of models whose usefulness is limited. As a consequence, not enough attention has been given to more general discussions about how to deal with climate change and to find alternative policy recommendations based on other approaches.

## 4.2    Pluralism in Economics and Schools of Thought

In the neoclassical tradition, climate change is seen as an externality problem that causes market failure. It is an externality problem of gigantic scale, but the basic logic of the problem is simple and explained in every introductory microeconomics textbook. In the published version of his Nobel Prize lecture, Nordhaus depicts the problem of climate change exactly in this way:

I begin with the fundamental problem posed by climate change—that it is a public good or externality. Such activities are ones whose costs or benefits spill outside the market and are not captured in market prices. ... Global warming is the most significant of all environmental externalities. It menaces our planet and looms over our future like a Colossus. (Nordhaus 2019, p. 1992)

Interpreting climate change as an externality alone trivialises the problem, since, from this point of view, the main research question boils down to determining the right carbon price. Once the optimal carbon price is found, it can be implemented by a tax and then the problem is solved, because the agents in the markets will internalise the social costs and partly avoid them. From this perspective, the problem of climate change is technically demanding, but, intellectually, rather boring. It is normal science in the Kuhnian sense, in which researchers do the routine work of solving technical problems. This routine work is rightly delegated to the field journals and only requires the attention of the experts, but not of the discipline as a whole.

Different schools of economic thought have different schemas to think about the world, to define research questions and to suggest approaches to answer the research questions. An economic school of thought is a specific way of thinking about economic issues shared by a community of researchers. There is a considerable number of non-mainstream schools of thought, such as ecological economics, feminist economics, or evolutionary economics, that frame the problem of climate change in a different way than neoclassical economics. One could imagine a state of economics, in which economists from different schools of thought discuss the different approaches and debate over the different policy conclusions implied by them. Economists from different camps could also join efforts and synthesise different approaches creating a basis for better policy recommendations. Dobusch and Kapeller (2014) call such a constructive engagement of different paradigms with the aim of "ecumenical integration and diversification" *interested pluralism*. However, what is actually practised is either *monism* or *selfish pluralism* or *disinterested pluralism*. Many mainstream economists have a monist conception of economics, meaning that there is just one correct, scientific way of practising economics, which is actually the neoclassical approach (Dusek 2008; Heise 2016). Consequently, mainstream economists mostly ignore writings of non-mainstream economists. Aistleitner et al. (2018) show that in the past 30 years more than 97% of all citations in 13 top mainstream journals refer to other mainstream journals and less than 3% to heterodox journals. In contrast, publications in 13 leading heterodox journals cite mainstream articles much more often, between 47.5% in the years 1989–2008 and 24.5% after 2008. However, Dobusch and Kapeller (2014) argue that many non-mainstream economists simply criticise mainstream papers and strive for dominance of their own approach, which would be a sign of selfish pluralism. The other widely taken stand is that of disinterested pluralism in which engagement with other schools takes place only occasionally and interactions with other scholars are mainly confined to the own niche.[1]

The distinction between selfish, interested and disinterested pluralism is helpful, but does not capture the rich discussion of the term pluralism. There is a discourse on the meaning and boundaries of pluralism in economics (Negru and Negru 2017), which we briefly sketch here.

Pluralism originates from the modern philosophy and was introduced as the opposite of monism. It developed over time and was transferred as a highly problematic concept to other social sciences (Breitling 1980). In economics, pluralism is associated with a methodological discontent within economics, the orthodox-heterodox debate, movements such as Rethinking Economics and calls for more tolerance and openness in teaching and practicing economics (Negru 2009). Dereniowska (in Reardon 2015, p. 276) expects "pluralism about pluralism" indicating that pluralism can have different forms, definitions and degrees. For her, the pluralism in economics is not so much a scientific question but a political question of ideology and power within the science and profession of economics. In a roundtable dialogue on pluralism in economics, different types were discussed, such as *methodological pluralism, theoretical pluralism, critical pluralism, reflexive pluralism, systematic pluralism* or *pedagogical pluralism*. Explaining the differences goes beyond the scope of this work. However, Negru and Negru (2017, p. 195) summarise that pluralism "means ways of recognising and accepting the variety of economic ideas and schools of thought in economics and in the economics curriculum. Pluralism represents an attitude of openness towards theories and models that are not necessarily heterodox". Nguyen (in Reardon 2015,

---

[1] In heterodox climate economics, this is not the case. There is a lively exchange between Post Keynesian macroeconomics, complexity economics, feminist economics, institutional economic and ecological economics. In fact, ecological economics itself is a platform for interdisciplinary exchange and very open to diverse approaches and disciplines.

p. 279) emphasises that "pluralism is about the willingness to engage in open, critical and constructive discussions with your opponents".

For our understanding of pluralism in the context of the book, tolerance and openness are key. Tolerance does exceed to accept the co-existence of a diversity of theories and approaches. It also requires showing respect for other scientists and their work. Openness includes the willingness to reflect on the own research and the diversity of reality but also to engage in the related discourse. This understanding of pluralism does not imply pure relativism or uncritical acceptance of methods and approaches in economics in the sense of an anything-goes attitude. Our proposition is that there should be interested pluralism in economics in order to deal adequately with climate change.

Pluralism is challenging for heterodox and orthodox economics. Heterodox economics is only pluralistic "if it is reflexive, and not anti-orthodox non-biased and not dogmatic, and it does not imply a pure relativism that is absolutized" (Negru and Negru 2017, p. 198). Although neoclassical economics could become pluralistic, when acknowledging the co-existence of paradigm in economics, we share the pessimism of Keen (in Reardon 2015, p. 283) who doubts this development.

Currently, mainstream climate economics is not pluralistic, but dominated by neoclassical thinking, which has two implications. Firstly, criticisms raised against neoclassical economics in general apply to mainstream climate economics, too. Secondly, every school of thought has a specific approach to look at the world. As Ha-Joon Chang (2014, p. 111) writes:

All theories ... necessarily involve abstraction and thus cannot capture every aspect of the complexity of the real world. This means that no theory is good at explaining everything. Each theory possesses particular strengths and weaknesses, depending on what it highlights and ignores, how it conceptualizes things and how it analyses relationships between them.

Thus, it is to be expected that neoclassical climate economics only focuses on some aspects of the relation between climate change and the economy, but neglects others, which may be equally or even more important.

We elaborate on these two implications in the following subsections and present what other economics schools of thought have to say on these issues. In our definition, an economic school of thought is a set of ideas, opinions and common practices that a community of researchers studying economic problems shares. We focus on nine different schools of thought:

1. *Austrian economics*
2. *Behavioural economics*
3. *Complexity economics*
4. *Ecological economics*
5. *Evolutionary economics*
6. *Feminist economics*
7. *Institutional economics*
8. *Marxian economics*
9. *Post-Keynesian economics.*

Ecological economics is a special case, because it is sometimes seen as a multi-paradigmatic platform that applies many different approaches to sustainability issues (Gowdy and Erickson 2005).

In general, economic schools of thought can be characterised by their ontology, their epistemology, their methodology and their dominant value statements.[2] We provide succinct characterisations of the schools of thought, emphasising those aspects we consider most relevant in the context of climate economics. In many practical cases, there are no clear demarcations whether a specific piece of research belongs to one school of thought or another. For practical work, such assignments are not necessary and, ultimately, all categorisations can be criticised. Nevertheless, they help us understand what researchers do and how they do it. Since all schools of thought overlap with others in some dimensions, we do not organise our discussion of climate economics along the schools. We rather first present each school very briefly, criticise neoclassical climate economics from multiple perspectives, and then present neglected topics relevant for climate economics emphasised by different schools of thought.

The starting points to define economic schools of thought are three ontological questions: (1) What is central economic object? (2) From which "thing" should economic inquiry start? (3) What fundamental assumptions are made about human beings? The main epistemological

---

[2] The collaborative open-source e-learning platform *Exploring Economics* provides a useful overview over various economic schools of thought and the dimensions along which they can be compared: https://www.exploring-economics.org/en/orientation. Our discussion is strongly inspired by this compilation.

question is whether social scientists can refer to ontologically real objects and to what extent interpretations matter. The different schools can be arranged on an epistemological spectrum between realism at one end and constructivism at the other. In terms of methodology, we can distinguish the predominance or acceptance of quantitative versus qualitative research methods, and whether the approach is formalistic and standardised or non-formalistic and idiosyncratic. Finally, the schools differ with regard to which value statements dominate and whether there is an implicit or explicit relationship to moral theories.

### 4.2.1   Austrian Economics

Austrian economists share some ontological convictions with neoclassical economists, especially the focus on scarcity as the core economic problem and the individual as the starting point of the inquiry (methodological individualism). The behavioural model of *homo oeconomicus* is also endorsed, but enriched and weakened by a role for social factors, such as institutions and power, and less focus on utility maximisation. In contrast to the neoclassical school, Austrian economists emphasise the importance of fundamental uncertainty and of change in the economy. A stark difference to the neoclassical approach is the Austrians' subscription to constructivism and strong subjectivism, which means that the world that matters is the one that is perceived by individuals. The aggregation and coordination of individuals' knowledge and action plans via markets is hence a core topic. The combination of subjectivism and the emphasis on fundamental uncertainty gives rise to a methodological position, which is strongly deductive, but also non-formalistic. The Austrian school is highly averse towards the use of mathematics and statistics in economics and favours verbal expositions of arguments and thought experiments. Many Austrians openly embrace positions from liberalism and value individual rights, in particular property rights, highly.

### 4.2.2   Behavioural Economics

There are two strands in behavioural economics which are often called "old" and "new" behavioural economics (see Sent 2004). Especially new behavioural economics is close to neoclassical economics and psychology, while old behavioural economics is further away from the neoclassical school and closer to sociology rather than psychology. Behavioural economics focuses on individuals and their behaviour, who are confronted with situations of choice—implying scarcity—and uncertainty. The core topic of behavioural economics is to replace the model of homo economicus by a more realistic model of man that takes into account bounded rationality, other-regarding preferences, emotions and other human "complications" (Mullainathan and Thaler 2000). Behavioural economists mostly take a realist position and conduct experiments inspired by the scientific method of the natural sciences. The methods have both deductive and inductive elements and are mainly formalistic (especially in new behavioural economics) with some less formal aspects, especially formal models of behaviour are tested against "plausible" alternatives. Thaler and Sunstein (2003) introduced *libertarian paternalism* as an ethical concept that gained much popularity in behavioural economics. It says that it is both possible and legitimate for private and public agents to influence the behaviour of individuals such that those make better choices. By exploiting insights from behavioural economics, benevolent third parties can help individuals to behave more rationally through a careful design of the *choice architecture*. Since agents are still free to choose, this approach is libertarian, although individuals are manipulated to perform certain actions.

### 4.2.3   Complexity Economics

Complexity economists view the economy as a complex adaptive system, in which many heterogeneous agents interact and adapt their behaviour to the behaviour of others and changes in their environment. From these interactions of individual agents, patterns at the meso and macro levels of the system emerge that are not easy or even impossible to predict by looking at the properties and the behaviour of the individuals alone. The emergent aggregate structures influence or constrain the actions of the agents, which is called *downward causation*. These systems are hence characterised by both permanent change and uncertainty. Complexity economics aims at understanding systems, but emphasises the interplay between different levels (micro, meso, macro) of a system and regards properties of systems as the result of the interactions of individuals. Individuals are boundedly rational and are not assumed to optimise objective functions due to the complexity of the problems. They are embedded in social contexts and learn or imitate the behaviour of others. Since complexity economists believe that systems are more than the sum of their parts, they are sceptical about the reductionist approach that tries to understand a system by ana-lysing its components in isolation. While they favour formal modelling that inevitably requires complexity reduction, they are aware of their lim-ited capability of understanding the full system. In non-linear systems, small changes can lead to very different outcomes, which make complexity economists cautious about postulating general economic laws. With regard to ideological statements, complexity economists reject simple market-versus-government views that either stress individual responsibility or collective responsibility for well-being. They emphasise that markets always depend on an institutional framework than can be affected by pub-lic policy. At the same time, they warn against direct government interven-tions due to often unpredictable side effects they might have in complex systems.

### 4.2.4  Ecological Economics

Ecological economics takes into account that the economy is embedded into society and the biosphere. While the economy is an open system with natural resources flowing into the economy and material waste flowing out of it again, the planet Earth is a closed system with regard to material flows. Since both natural resources and the absorptive capacity of sinks for waste are limited and the economy depends on material throughput, eco-nomic activity must be constrained by planetary boundaries. It follows that scarcity is a central problem in ecological economics. Climate change is also an important topic because economic growth takes us closer to the planetary boundaries and because society must be transformed in order to become more sustainable. Since many facts about the biosphere and the relation between the economy and the biosphere are (yet) unknown, uncertainty is a third major issue. Ecological economics focuses on systems and has a holistic view similar to the one of complexity economics. With regard to individuals, ecological economics includes insights from psy-chology in order to understand and potentially change cognitive and social processes. A crucial question is how preferences form and change since they are a reason for sustainable or unsustainable behaviour. Regarding epistemology, ecological economics assumes that there are real objects. Being informed by natural sciences, it is empirically orientated and (mod-erately) falsificationist. At the same time, many ecological economists have a post-normal understanding of science, meaning that they do not only want to understand the world but also see a responsibility to contribute to its transformations towards more sustainability. Acknowledging that social processes are driven by pluralistic and often conflicting values, they subscribe to transdisciplinary research, which means that science cannot provide solutions for society, but must develop acceptable improvements together with societal stakeholders. Methodologically, ecological economics is pluralistic and uses a host of both quantitative and qualitative research methods. Ecological economics takes a clear and open normative stand that sustainability should be an overarching goal of economic and political activity. Sustainability has a social dimension, which implies that intra- and intergenerational justice are important values. Ecological economics is also concerned with the question of what is a "good life" and how economic activity contributes to this end goal.

### 4.2.5    Evolutionary Economics

Similar to Austrian economics and ecological economics, evolutionary economics deals with questions of change, uncertainty and scarcity. The central topic is how the economy changes, especially due to technical and social innovation and the resulting processes of growth, development and structural change. Knowledge and its discovery are crucial topics, with knowledge being a real object. Since innovation implies true novelty, the future must be fundamentally uncertain such that optimal intertemporal planning is impossible and agents must be boundedly rational. Competition for scarce resources is important, but there can also be cooperation if it increases the fitness in the market and hence leads to success and survival. In line with complexity economics, evolutionary economics focuses on individuals whose interaction leads to emergent outcomes at the meso and macro levels. For evolutionary economists knowledge is a real object, but due to innovation it is always preliminary and potentially faulty. Overall knowledge is the emergent outcome of the individual, subjective components. In evolutionary economics, the research focus is on the emergence and diffusion of knowledge rather than on the issue which knowledge is true. Both formalistic and standardised methods, as well as more idiosyncratic qualitative methods, are used. Methods that allow the analysis of dynamic processes are favoured. The specific context of time and space is considered, and hence generalisation not always aspired to. Innovation and the creation of knowledge are valued positively, such that change and growth and policies that promote them also have a positive connotation.

### 4.2.6    Feminist Economics

Power relations and dominance are the central topic in feminist economics. Power imbalances within family and between genders are at the core, but other power relations, for example, related to ethnic or social groups are also important. They are a main driver of social and economic dynamics. Since power relations always involve several actors, feminist economics takes groups like the household as the starting point for inquiry. Behaviour is strongly influenced by gender, identity and social roles, which are socially constructed and context-dependent. Most feminist economists are rather critical with regard to the homo economicus. Economic phenomena occur in spheres that are often treated as separated, but in fact are necessarily connected, for instance, public and private or reproductive and productive. A key term in feminist epistemology is *situated knowledge*, which means that researchers are unavoidably embedded in a historical, cultural, social and economic context. This implies that the position of the researcher in society influences what she or he regards as relevant and which kind of research as scientific. There is a plurality of methodological approaches in feminist economics, including deductive and inductive thinking, formalist and non-formalist techniques, qualitative and quantitative methods. Feminist economics is comparable to ecological economics in that it also requests a transformation of the economy. Feminist economists demand reforms leading to less inequality, in particular, between genders and promoting emancipation.

### 4.2.7    Institutional Economics

As with behavioural economics, there is an old and a new version of institutional economics. The latter has its roots in the 1960s and 1970s and is closer to neoclassical economics than old institutional economics. The four Nobel laureates Ronald Coase, Douglass North, Elinor Ostrom and Oliver Williamson are associated with new institutional economics. Institutional economics is focused on understanding institutions, which can be defined as social rules that structure social interaction. These rules both make interactions in social context possible and constrain individual behaviour. The empowering and the constraining functions of institutions emphasise the importance of power in economic analysis. Institutions at different levels emerge from the interactions of individuals and change over time at different speeds. Understanding this social change is also a central topic. Similar to evolutionary economics, institutional economics is concerned with phenomena in real historical time and space and less interested in abstract general social laws. It hence favours precise descriptions of empirical phenomena and mid-range theories. There are both positivist and constructivist strands. Especially the older tradition is rather inductive, non-formalistic and willing to use qualitative research methods. New institutional economics is more formalistic. Almost by nature, institutional economics aim at producing socially useful knowledge and analyse how institutions can be altered or designed to improve society. While there is no uniform ideological tendency, there is a scepticism with regard to liberalism and unconstrained capitalism leading to a general openness towards social and political interventions in the economy.

### 4.2.8    Marxian Economics

Marxian economics is sometimes also called Marxian Political Economy, indicating that it provides an integrative analysis of the economy, society and politics. These systems are interdependent and coevolve in historical time. A main driver of the system dynamics is the struggle between the classes, that is, capital and labour, for power. Hence, change and dominance are the central questions. The unit of analysis are classes, that is, social groups. Social structure and material conditions determine the behaviour of individuals. At the systems level, Marxian economics assumes the existence of laws of motion that determine how the economy, society and the political system evolve. There are real causal mechanisms that generate empirical phenomena. However, these real causal mechanisms are not easy to observe. In the tradition of critical realism, Marxian economics hence rejects positivism as naive and argues that empirical correlations between events neither prove nor reject the existence of causal relationships between them. The researcher must use critical judgment whether a theory or empirical research allows making true statements about the generative mechanisms. Strong constructivism, on the other hand, is also rejected, because Marxian economists believe that a real world exists independent of the researcher, even though its perception by the researchers is influenced by the personal biography. Methodologically, Marxian economics is eclectic and uses formalistic methods, such as mathematical models, but also qualitative methods, for example, discourse analysis and detailed case studies. Due to its rejection of positivism, there is scepticism regarding econometric analyses. Marxian economics is explicitly not value-free, but normative and performative in that it aims at transforming society. Building on the critique of existing economics and social conditions, the goal is to fight dominance, exploitation and inequality in a broad sense. Since the main cause of inequality and dominance is seen in the structural conditions in capitalism, Marxian economists want to replace capitalism by a different economic system.

### 4.2.9    Post-Keynesian Economics

Fundamental uncertainty and dominance are at the centre of Post-Keynesian economics. The interesting units of analysis are classes or sectors rather than individuals. Much analysis takes place at the macroeconomic level of the economic system. There is conflict between classes like workers, capital owners and rentiers over income. Due to fundamental uncertainty, institutions play an important role in guiding individual behaviour. Hence the behaviour of individuals is characterised better by rules of thumb and social influence than by rational choice. The capitalist economic system is perceived to be dynamic and unstable, but still subject to certain regularities generated by causal mechanisms. Post-Keynesian economics takes a realist position and requires that assumptions and statements about the world should not contradict key aspects of reality. Economic theories should be both internally consistent—for example, respect accounting identities—but also overall coherent and in line with empirical observation. In contrast to neoclassical economics, predictive performance and deductive rigour are not seen as the main quality criteria of economic theories, since Post-Keynesian economists are aware of complexity and emergent outcomes. They favour realistic and relatively simple models that provide plausible explanations of economic phenomena and adequate description of their causal mechanisms. Formalistic and standardised research models, such as mathematical models and econometric analysis, are important in Post-Keynesian economics, but there is also room for non-formalistic logical reasoning, case studies and rich descriptions of institutions and behaviour. Ideologically, most Post-Keynesianisms are somewhere between liberalism and socialism and do not want to eliminate capitalism. However, they generally support the idea of making the capitalist system more stable and just, with levels of employment and low income inequality. There is a conviction that capitalism can and should be tamed by government interventions.

## 4.3    CRITIQUE OF MAINSTREAM NEOCLASSICAL CLIMATE ECONOMICS

The ontological, epistemological and methodological premises and normative foundations of neoclassical economics explain the weaknesses and blind spots of mainstream climate economics. We hence first characterise neoclassical economics along these dimensions before we apply the insights to climate economics in the tradition of Nordhaus. We will criticise neoclassical climate economics mainly from a complexity perspective, but many of these criticisms are compatible with other schools of thought, too.

### 4.3.1    Neoclassical Economics

In the ontological dimension, neoclassical economics differs markedly from the other schools of thought. The most distinctive feature is the exclusive focus on scarcity as the central economic question. In their introductory textbook, Samuelson and Nordhaus (2009, p. 4) write:

> Economics is the study of how societies use scarce resources to produce valuable goods and services and distribute them among different individuals.

Similarly, the textbook by Robert and Rubinfeld (2018, p. 26) says:

> Much of microeconomics is about limits—the limited incomes that consumers can spend on goods and services, the limited budgets and technical know-how that firms can use to produce things, and the limited number of hours in a week that workers can allocate to labor or leisure. But microeconomics is also about ways to make the most of these limits. More precisely, it is about the allocation of scarce resources.

Note that these (and other mainstream) authors do not qualify their definition to apply only to neoclassical (micro-)economics, but regard it as a general definition of economics. Other topics like dominance, change or uncertainty that are key to other schools of thought are not at the core of neoclassical economics.

As mentioned in the previous chapter, Arnsperger and Varoufakis (2006) argue that the paradigmatic core of neoclassical economics consists of the three axioms *methodological individualism, methodological instrumentalism* and *methodological equilibration*. Individuals are clearly the starting point of all analysis, which is reflected by the requirement that even all macroeconomic analyses need "rigorous microfoundations" (Wren-Lewis 2007, 2011 for a discussion). The second feature that distinguishes neoclassical economics from all other schools of thought is the atomist perspective on individuals. Individuals have their private preferences and make choices independent of other individuals and their social context. They make optimal choices given their preferences and the respective budget constraints. Importantly, neoclassical economics assume that preferences are fixed and characterised by *deep parameters* and observed changes in behaviour are to be explained by changes in the budget constraint, that is, prices and income, but not by preference changes. Every behaviour is interpreted as a conscious, deliberate choice among alternatives. In the same way as individuals are assumed to be isolated against social influences, the economy is separate from the larger social system and typically also from the natural environment. The economy is seen as a closed system without external impacts that could alter the system. Since the components of the economic system, for example, consumers and firms, are atomistic and independent of each other, their behaviour follows stable deterministic or probabilistic laws. The rational pursuit of self-interest and competition leads to equilibria in interactions, for instance, in markets. Hence markets are, apart from temporary random disturbances, assumed to be stable. Due to this inherent stability, uncertainty is not fundamental, as in many other schools of thought. Instead, neoclassical economics interprets uncertainty as *risk* that can be well described by known probability distributions over known events.

Neoclassical economists assume that there is an objective real world that can be studied independently from the observer by the use of the appropriate research tools in a similar way as natural systems can be analysed. It cannot be the aim of the researcher to describe reality in all its facets. The interest of neoclassical economists lies in general, universally applicable laws that can be uncovered by "the scientific method". According to Samuelson and Nordhaus (2009, p. 5)

> [e]conomists use the scientific approach to understand economic life. This involves observing economic affairs and drawing upon statistics and the historical record. ... Theoretical approaches allow economists to make broad generalizations ... In addition, economists have developed a specialized technique known as econometrics, which applies the tools of statistics to economic problems. Using econometrics, economists can sift through mountains of data to extract simple relationships.

In order to find these universal simple relationships, neoclassical economists look at *stylised facts*, which are broad generalisations that summarise data, and use abstraction, stylisation and simplification. They are disinterested in social, institutional, political or historical details, which are regarded as distracting peculiarities that clutter the underlying economic laws. The main criterion whether the stylisation, which may even be in conflict with observed facts, is valid is predictive power. Neoclassical economists like to compare themselves with classical physicists as the following quote from Robert and Rubinfeld (2018, p. 28) shows:

> No theory, whether in economics, physics, or any other science, is perfectly correct. The usefulness and validity of a theory depend on whether it succeeds in explaining and predicting the set of phenomena that it is intended to explain and predict. Theories, therefore, are continually tested against observation. As a result of this testing, they are often modified or refined and occasionally even discarded. The process of testing and refining theories is central to the development of economics as a science. When evaluating a theory, it is important to keep in mind that it is invariably imperfect. This is the case in every branch of science. In physics, for example, Boyle's law relates the volume, temperature, and pressure of a gas. The law is based on the assumption that individual molecules of a gas behave as though they were tiny, elastic billiard balls. Physicists today know that gas molecules do not, in fact, always behave like billiard balls, which is why Boyle's law breaks down under extremes of pressure and temperature. Under most conditions, however, it does an excellent job of predicting how the temperature of a gas will change when the pressure and volume change, and it is therefore an essential tool for engineers and scientists. The situation is much the same in economics. For example, because firms do not maximize their profits all the time, the theory of the firm has had only limited success in explaining certain aspects of firms' behavior, such as the timing of capital investment decisions. Nonetheless, the theory does explain a broad range of phenomena regarding the behavior, growth, and evolution of firms and industries, and has thus become an important tool for managers and policymakers.

Given this epistemological position, it is not surprising that deductive mathematical models and econometric techniques are the only acceptable research methods. Most other methods, with the exception of economic experiments, are seen as unscientific, subjective and unreliable. Likewise, neoclassical economists see themselves as neutral observers of the economy and insist that their research is value-free. There should be a clear separation of positive analysis, which neutrally describes causal relationships, from normative analysis involving value judgements. Neoclassical economists do not feel equipped with the appropriate instruments for normative analyses:

> There are no right or wrong answers to these questions because they involve ethics and values rather than facts. While economic analysis can inform these debates by examining the likely consequences of alternative policies, the answers can be resolved only by discussions and debates over society's fundamental values. (Samuelson and Nordhaus 2009, p. 6)

However, most would agree that inefficiency is unjust, because someone could be made better off without making someone else worse off.

### 4.3.2  Ontological Problems

From a complexity perspective, but also in line with evolutionary economics, Austrian economics, and ecological economics, neoclassical climate economics does not deal adequately with the dynamic and uncertain nature of the economy. DICE and other IAMs do look at economic growth, which is a dynamic phenomenon. However, they do so in a very special way, because they do not allow for an evolution of the system that involves transformation and structural change. The economic core of the DICE model is the aggregate production function (equation [4] in Chap. 3). While all variables in this equation depend on time and change over time, the function itself does not change. Two important drivers of growth—population growth and technological progress—are exogenous and the respective growth rates are assumed to be constant. Furthermore, climate change affects only output, but not the capital stock and the capital stock includes only produced capital, but not natural capital. All this is highly questionable in the context of climate change.

The point is not only that variables are missing or assumed to grow at constant rates, although there might be good arguments to assume that they are affected directly by climate change. From a complexity view, the more important point is that these models attempt to describe long-run aggregate relationships by stable functional forms that are unlikely to be stable, if there are major technological, environmental and social changes. According to Boulton et al. (2015), the future is the emergent and path-dependent outcome of a complex interplay between *patterns of relationships* and *events*. Processes at the micro and meso levels of the economy lead to emergent patterns at the macro level that appears robust and stable. The famous stylised facts of economic growth proposed by Kaldor (1957) can be seen as such patterns. They describe relationships between variables, for example, that output per worker and capital per worker grow at constant and identical rates over time, implying that the capital/output ratio is roughly constant over time. As any textbook on neoclassical growth theory explains, neoclassical growth models (of which DICE is a variant) are designed to match these stylised facts and to generate a *balanced-growth path*. While such patterns may be roughly stable for some time period, they

> are always under threat of disruption by events. These events might be major external shocks such as pandemics, tsunamis, wars, climate change or migration ... Events may also be internal variations from the norm—a higher-than-average birth or death rate, or the death or arrival of a significant leader. And of course what we regard as internal and external depends on how we draw the boundaries around the problem. (Boulton et al. 2015, p. 30)

The crucial ontological difference between neoclassical and complexity economics is that these events can change the aggregate relationships. In neoclassical DSGE models, events are captured by random shocks added to stable relationships. While they can cause temporary deviations of macroeconomic variables from the balanced growth path, they do not change the path and equilibrating mechanisms leading the system back on the balanced growth path over time. Complexity economics and all other schools of thought that emphasise the importance of change deny that the long-run evolution of the economy can be described and understood by models assuming the stability of aggregate patterns of relationships. If every event potentially can alter the patterns of relationships, every hypothetical path of the economy's evolution will be different, depending on the unique sequence of events. To make things even more complicated, many "events" that are relevant in the context of climate change are unlikely to be exogenous or random. The occurrence of heat waves, droughts, floodings, international conflicts or major migratory movements depends directly on economic and political decisions. When, where and how strongly abatements measures are undertaken will affect the likelihood of these events. If the future is path-dependent, the time-invariant damage function (5') that depends only on temperature does not make sense.

A related ontological problem is that neoclassical economics treats the economy as a closed system, which is separate from the rest of society. In the DICE model, there are almost no policy variables and there is no reference to any kind of institutions. The main policy variable is the emissions reduction rate $\mu$ that enters the abatement function $\Lambda$ (equation [6] in Chap. 3). Besides the consumption level, the emissions reduction rate is the central control variable of the model, but it is not explained how it is determined. The social planner chooses its level such that welfare is maximised subject to the respective constraints, but it remains unclear how this is achieved. Many public policy issues are introduced into the DICE model as exogenous assumptions or constraints. As discussed in Sect. 3.3, Nordhaus introduces exogenously given temperature targets into the model and uses the model to simulate the impact on the economy and the climate system. Many non-mainstream schools of thought would argue that both economic and climate conditions affect policy choices such that the political and societal responses should be endogenous. One might, for instance, assume that severe global warming could have an intentional or unintentional effect on productivity growth, either because society deliberately wants to slow down economic growth or because social tension and conflict inhibit research and development activities.

Finally, all non-mainstream schools of economic thought reject the atomistic model of economic agents, which is related to the refusal of treating the economy and society as separated systems. The atomistic agent shows up in the utility function (equation [2] in Chap. 3), in which utility is only obtained from individual consumption. There are no explicit social influences and the utility function has a fixed functional form with constant parameters. The only objective of individuals and society is the maximisation of this utility from consumption. Although Nordhaus interprets consumption in a broad sense, it is still directly coupled to produced output as shown in equation (7). The pursuit of what, for example, Maslow (1954) would call *psychological needs* and *self-fulfilment needs* or the *axiological needs* proposed by Max-Neef (1991), which depend crucially on the interaction with a social group and, in many cases, are unrelated to produced output, does not play a role. Furthermore, the parameter $\alpha$ in the utility function does not depend on social influences and is fixed over time. The same is true for the much debated rate of time preference $\rho$ in the discount factor $R$ (equation [3]). Assuming fixed preferences of an individual in the short run might be acceptable to explain everyday economic choices. For whole societies, however, it is undeniable that values and the cultural context of preferences change over the course of generations (see Inglehart and Welzel 2005). Value change might be of particular relevance for the analysis of environmental problems, since both private and political mitigation efforts might be a function of how society evaluates nature.

### 4.3.3   Epistemological Problems

Neoclassical economists believe in the existence of universally applicable economic laws. In order to discover them, abstraction and stylisation are justified and even necessary. The abstractions can go to the point where they are in open contrast to reality. The usefulness of this approach is often claimed without further proof by reference to the plausibility of unrealistic assumptions as reasonable first approximations and the predictive success of the approach. One of the main problems of this approach is that it is strongly driven by the aim of stating the economic laws in precise mathematical equations, for which often strong simplifying assumptions are necessary. Without simplifying assumptions, the mathematical models are not tractable and the desired "theorems" cannot be derived. But as we argue below, the simplifications are not harmless in many cases and lead to meaningless results.

The very core of DICE and similar models is the maximisation of the intertemporal welfare function (equation [1] in Chap. 3). The whole purpose of the model is to find the optimal, welfare-maximising price that should be imposed on carbon emissions in order to eliminate the externality problem. The approach of maximisation the welfare function is flawed for several reasons. Nordhaus and others use the conventional *social planner approach* to find the societal optimum. Since there is no global social planner, this method is defended by the argument that it is equivalent to a decentralised economy with intertemporally optimising households and profit maximising firms in a perfectly competitive market equilibrium. Yet these conditions are not met either.

There is considerable evidence from behavioural economics that the conventional *discounted utility model*, which is the core of the Ramsey growth model, is not a good description of households' intertemporal consumption choice. Frederick et al. (2004, p. 201) conclude their review article as follows:

The DU model, which continues to be widely used by economists, has little empirical support. Even its developers—Samuelson who originally proposed the model, and Koopmans, who provided the first axiomatic derivation— had concerns about its descriptive realism, and it was never empirically validated as the appropriate model for intertemporal choice. Indeed, virtually every core and ancillary assumption of the DU model has been called into question by empirical evidence collected in the past two decades.

In addition, consumers are heterogeneous and differ considerably in their planning behaviour (Ameriks et al. 2003; van Rooij et al. 2012) and in their desire to use a maximising or a satisfying decision-making style (Parker et al. 2007). Due to the heterogeneity of consumers, it is not surprising that the empirical support for the neoclassical model of intertemporal consumption choice is rather weak (D'Orlando and Sanfilippo 2010).

In a similar vein, complexity economists reject the assumption that firms maximise profits (Beinhocker 2007). While sufficient profitability is a necessary condition for firms to endure, the objectives of firms are multidimensional and include the interest of many different stakeholders apart from the owners. Post-Keynesian economists share the view that the management of firms pursues multiple objectives, of which firm survival is an important one (Lavoie 2006). In order to survive, firms have to grow and to acquire power, for example, by larger market shares. Finally, evolutionary economists in the tradition of Nelson and Winter (1982) would also agree that firms do seek profits, but that the maximisation of profits is impossible due to uncertainty, bounded rationality and organisational complexity. A related point is that most non-mainstream schools reject the assumption of perfect competition and price-taking behaviour as an adequate description of the real world. Firms permanently strive for monopoly and price-setting power and many markets are oligopolies with rather small numbers of strategic competitors. In most countries, there are also various forms of state intervention in markets that prevent perfect competition.

If neither households nor firms maximise their objective functions and if markets are not characterised well by a perfectly competitive equilibrium, the market outcome will most likely be different from what a social planner would have done in order to maximise social welfare. But then climate models using the social planner approach are purely theoretical exercises with little external validity. One might argue that models built on these stylised and unrealistic assumptions still capture many observed facts of the real economy. Even if this were true, there is no basis for the claim that the market outcome is maximising societal welfare.

A final problem applies to the interpretation of the utility and the welfare function. Nordhaus (2013) stresses that the utility levels of each period concern the respective generation or cohort of people on earth. Hence, the welfare function is a social welfare function, which relates the welfare of different generations to one another over time. As a consequence, he does not consider the key parameters $\alpha$ and $\rho$ as psychological characteristics of individuals but rather as social parameters that refer to large groups of people. This interpretation is in contrast to the usual interpretation of the utility function describing the utility of the *representative agent*, in which an explicit connection between the aggregate social level and the individual level is made. In line with textbook treatments and the usual approach in the literature, Nordhaus (1992, p. 1315) says that "the DICE model treats the world as a single economic entity and analyses the optimal policy for the average individual". In the representative agent framework, it is assumed that the utility function describes the preferences of individuals such that the parameters of the utility function are psychological traits of persons that could in principle be measured. The individual preferences are then aggregated into social preferences of society assuming that the social preferences are equal to the preferences of the "average individual". Although Kirman (1992), Hartley (1997) and others have shown long ago that the use of representative agents in macroeconomics is fundamentally flawed, they are still widely used in macroeconomic models. Blanchard (2009, p. 208) claims that "the state of macro is good" and the main merit of macro models with microfoundations is that welfare analysis with the correct welfare criterion is possible. Kirman (1992) demonstrates with a simple example that the interpretation of the representative agent as an average of different individuals does not make sense, because aggregating the revealed preferences of two individuals between two options into the choice of one representative agent can lead to the opposite choice of the representative agent relative to the choices of both individuals. Hartley (1997) shows that the representative agent constructed from individual choices in general is not stable if there is income redistribution. Only if all individuals have identical homothetic preferences generating parallel linear Engel curves, consistent aggregation is possible. Furthermore, they must have identical incomes in order to guarantee the existence and uniqueness of a general equilibrium (Kirman 1992). Assuming that all individuals have very specific and identical utility functions and identical incomes is clearly counterfactual. Yet, if the utility function and the welfare function are not the result of the aggregated individual's preference, it is unclear what the welfare levels of generations might be and how they could be measured. Nordhaus (2013) tries to argue that a *descriptive approach* of determining the social preference parameters could be used. Under the assumptions of utility and profit maximisation in a competitive equilibrium, the *Ramsey Equation* should hold:

$$r^* = \rho + \alpha g^*. \tag{18}$$

The Ramsey Equation relates the equilibrium real interest rate $r^*$ to the equilibrium long-run growth rate $g^*$ and the two preference parameters. Even if the two equilibrium levels of the interest rate and the growth rate could be observed, which is questionable, one of the two parameters remains undetermined. As Nordhaus (2013) demonstrates himself, the choice of the two parameters has a strong impact on the estimated social cost of carbon. Because there is no objective way of measuring or estimating the free parameter of the hypothetical social welfare function, the neoclassical approach of climate economics does not meet its own epistemological requirement according to which "[t]heories … are continually tested against observation. … The process of testing and refining theories is central to the development of economics as a science" (Robert and Rubinfeld 2018, p. 28).

Because of the discussed epistemological problems of the neoclassical approach, we doubt that mainstream climate economics can achieve its goal of deriving optimal policies to respond to climate change in a meaningful way.

### 4.3.4   Methodological Problems

On the methodological side, neoclassical climate economics is not only criticised from non-mainstream positions but also from within the mainstream. Pindyck (2013, 2017) argues that IAMs like DICE are not useful for policymaking and can even be misleading. His main argument is that these models provide a "veneer of scientific legitimacy" to policy recommendations that are driven by researchers' beliefs rather than objective scientific facts. Pindyck makes the point that in IAMs crucial parameters, such as the discount factor and the climate sensitivity parameter, and functional forms, for example, of the damage function, are arbitrary and cannot be known. Researchers have no way to find the right parameter values or functional forms empirically or theoretically. This is a problem, because different assumptions about crucial parameters like the discount factor have a tremendous impact on the results of the models. Depending on the subjective beliefs of researchers about the appropriate discount factor or climate sensitivity, the social costs of carbon can be very low or very high. The problem is not that researchers have different opinions or make value judgments on how to deal with the discount factor. The problem is that complicated models are like black boxes and obfuscate that the policy conclusions are strongly driven by these opinions. Pindyck criticises that the validity of these models is oversold and scientific honesty would require researchers to state that they do not know the key parameters. He also opposes the common view of many mainstream economists that a flawed model is better than no formal model at all. For him, surveys among experts about their beliefs on the social cost of carbon are informative enough for policymaking and more transparent than the IAMs (Pindyck 2019).

Weitzman (2009) presents his *Dismal Theorem* stating that the existence of small but positive probabilities of catastrophic effects of climate change renders the climate problem unsuitable for cost-benefit analysis. Assuming thin-tailed distributions such as the normal when in fact there is a fat-tail risk of catastrophic outcomes produces misleading conclusions. In fact, the consequences of the structural uncertainty about tail-risks are much more significant than the widely discussed effects of the appropriate choice of the discount factor. Like Pindyck, Weitzman warns against using certain methods to give one's analysis a more scientific appearance.

### 4.3.5   Problems with Normativity

No other school of thought insists as strongly as neoclassical economics that its research is value-free and objective. The feminist economist Julie Nelson (2008) explains this with the masculine self-image of scientists. She calls the belief that researchers could take a perspective-free viewpoint on their object of study *objectivism* and argues that it is impossible to derive perfectly objective knowledge because "scientists—and economists—are inherently embedded in nature, embedded in society, and hence part of, and inherently interested in, the very phenomena we study" (Nelson 2008, p. 443).

The most obvious tension between normativity and neoclassical climate economists' pursuit of objectivity concerns the discussion about the appropriate discount rate in the IAMs. The famous *Stern Review on the Economic of Climate Change* (Stern 2007) criticised Nordhaus for using discount rates that are too high. Stern openly argued that the choice of the discount rate is an ethical one and that justice towards the future generation requires very low or even no discounting at all. Nordhaus (2007, 2013), however, rejects what he calls the *prescriptive view* that makes ethical arguments and prefers the *descriptive approach*, according to which the discount factor should be chosen to reflect market realities. He states:

> This approach assumes that investments to slow climate change must compete with investments in other areas. The benchmark … should therefore reflect the opportunity cost of investment. …It is inefficient, in the descriptive view, to accept investments in climate mitigation with a yield of 1.4% per year if there are available investments in education or capital with yields of 6% per year. (Nordhaus 2013, p. 1114)

As Nelson (2008) and many others explain, the apparently objective approach is not value-free, but just hides ethical assumptions. It is clearly a value judgement that climate policy should be *optimal* or *efficient*. An alternative value judgement could require that it should be *effective* or *precautionary* in the sense that it prevents catastrophic outcomes with a high likelihood.

More generally, value judgements are present in many modelling choices. It seems hard to argue that a social welfare function can be chosen in a value-free way. Apart from the problem of intergenerational justice that is affected by the choice of the discount factor, a representative agent approach that does not allow talking about distributional issues within a generation is also a value judgement. Focusing on the average agent (if the representative agent could be interpreted in that way, which it is not possible) makes it impossible to ask whether climate change affects rich and poor agents in the economy or skilled and unskilled workers or other distinguishable groups in different ways. It also leaves out the question who has to contribute how much to mitigation efforts, which is a key political issue. Choosing a modelling framework that cannot deal with distributional effects on heterogeneous agents is an implicit statement that these issues are less important than the aspects that one can analyse with the model. This modelling choice hence is a value judgement. Steigleder (2018) argues that including climate-related health damages and even fatalities into the damage function and thus including them in a cost-benefit analysis is an unacceptable normative position. Rights-based moral theories judge actions or policy with regard to their consequences on the rights of the persons that are affected. Furthermore, they assume that there is a hierarchy of rights with one person's right to live being more important than another person's right to consume more goods and services. If a neoclassical climate economist implicitly concludes that the consumption benefits of some people due to slightly less mitigation efforts outweigh the costs in terms of additional deaths of some people due to higher temperature, he clearly makes a value judgement. Yet this value judgement is hidden and not easy to detect in the mathematical language.

At the ontological level, value judgements are present, too. According to Joseph Schumpeter (1954) every economic analysis is preceded by a preanalytic cognitive act, by which the researcher visualises "a distinct set of coherent phenomena as a worth-while object of ... analytic efforts" (pp. 38–39). Schumpeter calls this preanalytic cognitive act the *Vision*, which "supplies the raw material for the analytic effort" (p. 39). The Vision cannot be value-free:

> Now it should be perfectly clear that there is a wide gate for ideology to enter into this process. In fact, it enters on the very ground floor, into the preanalytic cognitive act of which we have been speaking. Analytic work begins with material provided by our vision of things, and this vision is ideological almost by definition. It embodies the picture of things as we see them, and wherever there is any possible motive for wishing to see them in a given rather than another light, the way in which we see things can hardly be distinguished from the way in which we wish to see them. (Schumpeter 1954, p. 40)

Schumpeter's Vision is related to what is called *mental model* in the psychological theory of reasoning by Johnson-Laird (1983) and Johnson-Laird and Byrne (1991). Mental models are information filters that cause selective perception only of some information about the world. The value theory by the social psychologist Shalom Schwartz (Schwartz 1992, 1994) postulates that

[v]alues serve as standards or criteria. Values guide the selection or evaluation of actions, policies, people, and events. People decide what is good or bad, justified or illegitimate, worth doing or avoiding, based on possible consequences for their cherished values. But the impact of values in everyday decisions is rarely conscious. Values enter awareness when the actions or judgments one is considering have conflicting implications for different values one cherishes. (Schwartz 2012, p. 4)

The claim that climate economics or any other scientific endeavour could be value-free hence contradicts both reasoning in philosophy of science and psychological research on the functioning of the human mind (including that of researchers).

## 4.4 Implications for Climate Policy

Since the neoclassical approach frames global warming as a problem of market failure, the standard solution is to heal the market by correcting the prices. Yet this solution is neither the only one which is possible nor necessarily the best. Some philosophers forcefully argue that there are moral limits of markets (Satz 2010; Sandel 2012). Given that global warming caused by greenhouse gas emissions can have widespread and serious consequences, a "right" to pollute the atmosphere may be hard to justify. In addition to ethical considerations about markets as universal instruments of coordinating and governing human behaviour, there is economic research arguing that human behaviour is different depending on whether a situation is framed in a market context or a non-market context. In particular, a market framing may drive out the intrinsic motivation to do something and replace it by extrinsic motivation (Bowles 1998; Gneezy and Rusticchini 2000a; Frey and Jegen 2001). Within a market context, behaviour is guided by competition and incentives, whereas it is often guided by norms of cooperation in non-market situations (Vatn 2009). Neoclassical economists typically argue that societal problems must be tackled by setting appropriate incentives to induce the desired behaviour, but not by moral preaching (Steckel and Peters 2019). However, Gneezy and Rusticchini (2000b) have shown experimentally that introducing incentives can actually reinforce undesired behaviour. In a day-care centre for children in Haifa, parents often picked up their children too late. Introducing a small fine for being late led to even more tardiness, because parents interpreted the fine as a price they could pay for more convenience. While it is surely correct that climate change is a problem that cannot be solved at the individual level alone and that decisive government action is needed, it is not so clear that the only policy must be to impose the right market incentives in the form of prices and subsidies. Other instruments such as direct regulation by laws may be needed as well.

Mainstream climate economics focuses too much on the question of the "right" carbon price. We do not argue that carbon pricing is wrong; it might be one important element in mitigation strategies. However, the question of its optimal level seems to be of secondary importance and cannot be answered in a scientific way or at least not using IAMs. The DICE model and similar approaches leave many important questions unanswered that are highly relevant for understanding the relation between global warming and the economy and for the design of effective mitigation and adaptation policies.

We suggest that climate economics should address at least six topics that have not received enough attention in the economic mainstream so far:

1. *How could effective mitigation policies be implemented?*
2. *What are the necessary conditions for technological countermeasures?*
3. *How could the behaviour of people become more sustainable?*
4. *What are the detailed consequences of global warming on economies and societies?*
5. *What are the factors that determine actual economic and political decision-making relevant for climate change?*
6. *Is it necessary to change the economic and political system and if so, how?*

In the following subsections, we briefly discuss these topics and point to what non-mainstream schools of thought could contribute to their analysis.

### 4.4.1    Implementation of Climate Policies

Nordhaus (2019, p. 1992) argues that "[g]overnance is a central issue in dealing with global externalities because effective management requires the concerted action of major countries" and that getting to international agreements on a global carbon price is the greatest challenge in climate economics. Paavola and Adger (2005) agree with this assessment and suggest that institutional economics can be an important source of ideas how governance problem could be understood and how climate policies could be designed. They argue that Coase (1960) started new institutional economics as an alternative intellectual research programme and as a critical response to the Pigouvian treatment of externalities. A key point of the institutional critics is that the separation of allocative decisions from distributional issues that is standard in the externality framework is illegitimate. Environmental problems should be interpreted as instances of *interdependence* instead of externalities. Interdependence means that the choices of one agent affect the choices of another. The problem is that "[i]nterdependent agents cannot simultaneously realise their incompatible interests in scarce environmental resources and their conflict must be resolved by defining (or re-defining) initial endowments" (Paavola and Adger 2005, p. 355). Governance institutions, which can be informal or formal and range from the local to the international scale, have the function to resolve such environmental conflicts. Environmental governance deals with issues of establishing, transforming and enforcing governance institutions such as environmental agencies or collective agreements.

The main message of the institutional critique of mainstream climate economics is that the research focus should not be on the policy instrument (i.e. the carbon price), but on how policies could be implemented against the backdrop that there is interdependence of relevant agents and conflicting interests. The processes and procedures by which environmental decisions are made should receive more attention. For the implementation of policies, the perceived distributive and procedural justice is more important than the overall welfare implications. Paavola and Adger (2005) present an overview over relevant literature and call for *institutional ecological economics* as a field with an important and innovative research agenda.

### 4.4.2 Conditions for the Successful Adoption of Decarbonisation Technologies

The neoclassical logic of the internalisation of external costs due to a Pigouvian tax says that profit-maximising firms will avoid $CO_2$ emissions if the previously free emissions get a price. One way to avoid $CO_2$ emissions is by investing into less carbon intensive technologies or in technologies that capture $CO_2$ after the process that generated the emissions.

Van den Bergh (2007) explains why this neoclassical view is not sufficient from the perspective of evolutionary economics. Neoclassical economics does not analyse how innovation occurs and what might be factors that promote or hinder the development of new less carbon-intensive technologies. A particular problem is that economy-wide decarbonisation requires not only the adoption of new technologies by individual firms but the transformation of whole systems, such as the energy system, the transportation system or the agricultural production system. The transformation of systems is difficult because of technological complementarities, supply-chains, path dependence and lock-in effects. For example, the transition from gasoline-fuelled cars to electric cars or hydrogen-fuelled ones cannot simply be achieved by a decision of car manufacturers. Customers will not buy such cars as long as there is no infrastructure for fuelling hydrogen cars or recharging electric cars. Furthermore, enough electricity or hydrogen must be produced by someone. These system transitions involve significant coordination problems by a multitude of different actors and are hence impossible to pursue by individual firms, even if they have a monetary incentive to do so. Van den Berg (2007, p. 540) states:

> From the perspective of path-dependence and lock-in, relevant questions are how regime shifts occur, how they can be stimulated, and how new lock-ins of inefficient or undesired technologies can be avoided. Preventing early lock-in requires a sort of portfolio investment. Un-locking of undesired structures and technologies—from an environmental or some social welfare perspective—cannot be realized by simply 'correcting prices', but requires also taking into account increasing returns on demand and supply sides, and (potential) learning curve effects. This has been referred to as transition policy … and it requires additional policies, notably innovation policy.

Given that there exists uncertainty about the choice of the right technology, evolutionary economics emphasises that policy strategies should aim at reducing risk and generating diversity and adaptive flexibility in the system. Diversity and learning are central in evolutionary processes to find solutions that are well-adapted to complex environments. From a neoclassical perspective, fostering diversity appears inefficient. In line with our previous discussion, van den Bergh (2007, p. 541) argues that the neoclassical focus on optimality is misguided:

> The traditional economic theory of environmental policy is the result of applying neoclassical welfare theory, which comes down to connecting a competitive equilibrium and a social welfare optimum (or, less ambitiously, Pareto efficient situation). Evolutionary systems based on bounded rationality, non-equilibrium and path-dependence offer a less optimistic view on policy, and imply that the normative part of neoclassical economics is based

on partly incorrect assumptions. If, due to this, the correspondence between a market equilibrium and a social welfare optimum is lost, then inevitably the two fundamental theorems of welfare economics no longer hold. As a result, planning or market solutions cannot be guaranteed to deliver socially optimal outcomes.

Evolutionary economics offers a different perspective on the role of policy in the context of climate change. Rather than just providing the incentive to reduce carbon emissions, it must help overcome coordination problems and must make the system innovative, adaptive and cooperative by innovation policy, regulation and the creation of appropriate institutions.

### 4.4.3    Sustainable Behaviour

The neoclassical rationale for carbon prices with regard to households is the same as for firms. If consumers are faced with the true costs of their consumption choices, they are believed to reduce carbon-intensive consumption and choose more climate-friendly goods and services. This mechanism from neoclassical consumer theory is flawed for at least three reasons.

First, there might not exist any appropriate climate-friendly substitutes for the more expensive carbon-intensive goods. If consumers rely on their cars for commuting, they cannot simply switch as long as either adequate public transportation or cars with alternative engines are available. Making gasoline more expensive just reduces their income, at least in the short run. There is co-evolution between consumer behaviour and the supply of goods and services by firms. Second, research in behavioural economics suggests that the responses of consumers to higher prices due to carbon taxes might be much weaker than predicted by neoclassical theory. Congdon et al. (2009) argue that consumers can respond to taxes in unexpected ways because of imperfect rationality or non-standard preferences. Gowdy (2008) adds that behaviour does not only depend on relative prices. Monetary incentives can crowd out other motivations causing perverse effects. Third, in neoclassical climate economics, carbon prices are exogenously imposed on the system by the social planner. However, policymakers in democratic societies need political support for levying new taxes. If there is no majority in favour of climate protection policies, then the government will not implement them. In other words, climate policy is endogenous and does not only influence consumer choices but also depends on the actions of consumers that are also voters.

Scholars from ecological economics focus on how the preferences and values of consumers and voters change. In contrast to the neoclassical assumption of fixed preferences, ecological economics and behavioural economics (Fehr and Hoff 2011) allow for social learning and changing preferences and values. Understanding how values change and how policies and institutions could be designed to facilitate social learning is a key aspect emphasised by ecological economics as a branch of sustainability science (Miller et al. 2014).

### 4.4.4   Understanding the Effects of Climate Change on the Economy

From a complexity perspective modelling the social world as if it would not change fundamentally in response to climate change is an important flaw of neoclassical climate economics. As Arthur (2015, p. 1) puts it:

> We also see the economy not as something given and existing but forming from a constantly developing set of technological innovations, institutions, and arrangements. Complexity economics thus sees the economy as in motion, perpetually 'computing' itself—perpetually constructing itself anew. Where equilibrium economics emphasizes order, determinacy, deduction, and stasis, complexity economics emphasizes contingency, indeterminacy, sense-making, and openness to change.

Climate-induced change of society could lead to very different futures. Using the method of qualitative scenario development, Raskin et al. (2002) distinguish three different kinds of futures: (1) Conventional Worlds, (2) Barbarization, and (3) Great Transitions. Conventional worlds are futures that are similar to the present and marked by adjustments to climate change either induced by markets or by policy reforms. In contrast, great transitions involve major, actively pursued transformations of society and the economic system. The barbarisation scenarios suggest that there could also be a breakdown of societal order with conflict and institutional disintegration. The possibility of societal collapse is also discussed in political science (Brunk 2002a), geography (Diamond 2005), archaeology (Tainter 1988), physics (Livni 2019), sustainability science (Abel et al. 2006), and Marxian economics (Foster 2016). Historically, many civilisations such as the Roman Empire or the Mayan Civilisation collapsed due to a combination of external triggers and complex internal dynamics. The concept of *self-organised criticality* from complexity science postulates that systems that developed more and more complexity and have been stable for considerable time, very suddenly can become unstable due to sometimes trivial and small events (Brunk 2002b). Against this backdrop, some scholars discuss whether the concept of sustainability should be replaced by the concept of *resilience* in the field of socio-ecological analysis (Davidson 2010; Benson and Craig 2014). The flip side of resilience is *vulnerability*, that is, the inability of a system to withstand external shocks.

The ideas of vulnerability and resilience of social systems imply that climate economics should pay more attention to two issues. First, a detailed investigation is needed of how climate change might alter the social environment in which economic activity takes place. For instance, climate change is likely to induce large-scale regional and international migration, which does not only affect labour markets, but also the social and political system. Political resistance against immigration might lead to a reversion of economic globalisation with structural implications for open economies and lower growth prospects. Second, climate economics should put less emphasis on forecasting the likely effects of climate change and more on foresight or anticipation of possible effects (Boyd et al. 2015; Roos 2016, 2018). The problem with forecasts is that they are almost always wrong, especially if the forecast horizon reaches far into the future. Policies to increase the resilience of a system, which are based on most likely forecasts, will only work well if the forecast is realised, but might perform poorly if the forecast was wrong. Anticipating possible futures instead of predicting the single most likely one allows policymakers to take precautionary actions for different scenarios. Being prepared for several states of the world instead of just one lowers the vulnerability against external events that might appear unlikely.

### 4.4.5 Political Economy and Climate Change

The implementation of mitigation and adaptation policies does not only depend on conflict-resolving institutions, but also on the political process or the politics of climate change. Tanner and Allouche (2011, p. 2) define *the political economy* "as the processes by which ideas, power and resources are conceptualised, negotiated and implemented by different groups at different states". While the issue of conflict and power relations overlaps with the interest of institutional economics, political economy puts an emphasis on the ideas and ideologies of the involved groups. They are important because they define whether there is a problem that should be tackled by society and the government and how this should be done. Levy and Spicer (2013, p. 660) speak of *imaginaries* that

> provide a shared sense of meaning, coherence and orientation around highly complex issues. Imaginaries are closely linked to the ways in which institutions and economic activity are organised and structured and to the ways people think they ought to be organised and structured.

They analyse how NGOs, business and state agencies contested four climate imaginaries in the US since the 1990s: 'fossil fuels forever', 'climate apocalypse', 'techno-market' and 'sustainable lifestyles'. In politics, the different groups and parties aim at making their own imaginary or worldview the dominant one in order to gain control over societal resources and the decisions how they should be used.

The neoclassical view on climate policy is apolitical and has an apolitical understanding of the policy process. It assumes a linear policy process in which a problem is first identified by the government or science and then tackled in an orderly process governed by economic rationality. Real political processes, however, are messy and characterised by ambiguity. Political goals are often not clearly defined and neither are the problems. According to Tanner and Allouche (2011, p. 11):

> the growing importance of climate change in the development arena and the frequent assumption of linear policymaking and apolitical, techno-managerial solutions make the development of a new political economy emphasis vital to determining efficient, equitable and effective responses. ... [W]e suggest that explicit attention is given to the way that ideas, power and resources are conceptualised, negotiated and implemented by different groups at different scales.

### 4.4.6   Rethinking of Capitalism and System Change

A key issue in ecological economics is whether permanent economic growth and sustainability can be reconciled. Ecological economists distinguish *absolute* and *relative decoupling*. Relative decoupling means that the resource intensity per unit of produced output declines, which means that the total resource impact of production still can rise if the economy grows.

Therefore, relative decoupling does not solve the problem of hitting the planetary boundaries at some point in time if there is ongoing growth. Some ecological economists hence conclude that either absolute decoupling is necessary, that is, an absolute decrease of resource impact, or the economy must stop growing. Since they doubt that absolute decoupling is possible, they favour the idea of the *steady-state economy* (see Daly 1977; Jackson 2009) which does not grow.

The steady-state or post-growth literature raises two important questions. The first question is about the purpose of economic growth and economic activity in general. Skidelsky and Skidelsky (2012) argue that economic activity should be a means to the end of a good life, but not an end in itself. But this raises the question of what makes a good life and which role employment, consumption and material wealth play. These reflexions often imply a critique of consumerism and capitalism, that is, of the current economic system in Western countries, because this system sets the wrong priorities. The second question is whether capitalism can be reformed to become more sustainable or whether this is not possible. Marxian economists such as Foster et al. (2016) argue that capitalism should be overcome in an ecological revolution and replaced by a system that allows sustainable development. Others debate whether there is a growth imperative in capitalist market economies or not (Gordon and Rosenthal 2003; Binswanger 2016; Richters and Siemoneit 2017, 2019). If there was no growth imperative, it might be possible to reform capitalism in order to reach post-growth market economies.

Neoclassical climate economics shuns these fundamental questions about the role of economic activity in society and the life of people and how it could and should be organised. Yet they are of prime importance, because if the current economic system is fundamentally unsustainable, tinkering with policy instruments such as a carbon tax without major transformations of the system might not avoid climate disasters.

## REFERENCES

Abel, Nick, David H.M. Cumming, and John M. Anderies. 2006. Collapse and Reorganization in Social-Ecological Systems: Questions, Some Ideas, and Policy Implications. *Ecology and Society* 11 (1): 17.

Aistleitner, Matthias, Jakob Kapeller, and Stefan Steinerberger. 2018. The Power of Scientometrics and the Development of Economics. *Journal of Economic Issues* 52 (3): 816–834.

Ameriks, John, Andrew Caplin, and John Leahy. 2003. Wealth accumulation and the propensity to plan. *Quarterly Journal of Economics* 118 (3): 1007–1048.

Arnsperger, Christian, and Yanis Varoufakis. 2006. What is Neoclassical Economics?: The Three Axioms Responsible for its Theoretical Oeuvre, Practical Irrelevance and, thus, Discursive Power. *Panoeconomicus* 53 (1): 5–18.

Arthur, W.B. 2015. *Complexity and the Economy*. Oxford: Oxford University Press.

Beinhocker, E.D. 2007. *The origin of wealth: The radical remaking of economics and what it means for business and society*. Harvard Business Press.

Benson, Melinda Harm, and Robin Kundis, Craig. 2014. The End of Sustainability. *Society & Natural Resources* 27 (7): 777–782.

Binswanger, Mathias. 2009. Is there a growth imperative in capitalist economies? A circular flow perspective. *Journal of Post Keynesian Economics* 31 (4): 707–727.

Blanchard, Olivier J. 2009. The State of Macro. *Annual Review of Economics* 1: 209–228.

Boulton, Jean G., Allen, Peter M. and Bowman, Cliff. 2015. *Embracing Complexity – Strategic Perspectives for an Age of Turbulence*. Oxford: Oxford University Press.

Bowles, Samuel. 1998. Endogenous preferences: The cultural consequences of markets and other economic institutions. *Journal of economic literature* 36 (1): 75–111.

Boyd, Emily, Björn Nykvist, Sara Borgström, and Izabela A. Stacewicz. 2015. Anticipatory Governance for Social-Ecological Resilience. *Ambio* 44 (Suppl. 1): 149–161.

Breitling, Rupert. 1980. The Concept of Pluralism. In *Three Faces of Pluralism*, ed. Sergej Ehrlich and Graham Wootton, 1–19. Westmead: Gowe ver.

Brunk, Gregory G. 2002a. Why Do Societies Collapse? A Theory Based on Self-Organized Criticality. *Journal of Theoretical Politics* 14 (2): 195–230.

———. 2002b. Why Are So Many Important Events Unpredictable? Self-Organized Criticality as the 'Engine of History'. *Japanese Journal of Political Science* 3 (1): 25–44.

Chang, Ha-Joon. 2014. *Economics: The User's Guide*. London: Penguin Books.

Coase, R.H. 1960. The Problem of Social Cost. *The Journal of Law and Economics* 3: 1–44.

Congdon, William, Jeffrey R. Kling, and Sendhil Mullainathan. 2009. Behavioral Economics and Tax Policy. No. w15328. National Bureau of Economic Research.

D'Orlando, Fabio, and Eleonora Sanfilippo. 2010. Behavioral Foundations for the Keynesian Consumption Function. *Journal of Economic Psychology* 31 (6): 1035–1046.

Daly, Herman. 1977. *Steady-State Economics*. Washington, DC: Island Press.

Davidson, Debra J. (2010). The applicability of the concept of resilience to social systems: some sources of optimism and nagging doubts. *Society & Natural Resources: An International Journal* 23 (12): 1135–1149.

Diamond, Jared. 2005. *Collapse: How Societies Choose to Fail or Succeed*. London: Penguin Books.

Dobusch, L. and Kapeller, J. 2014. Heterodox United vs. Mainstream City? Sketching a Framework for Interested Pluralism in Economics. *Journal of Economic Issues* 46 (4): 1035–1058.

Dusek, Tamás. 2008. Methodological Monism in Economics. *Journal of Philosophical Economics I* 2: 26–50.

Fehr, Erst, and Karla Hoff. 2011. Introduction: Tastes, Castes and Culture: The Influence of Society on Preferences. *Economic Journal* 121 (Nov.): F396–F412.

Foster, John Bellamy. 2016. The Earth-System Crisis and Ecological Civilization: A Marxian View. *International Critical Thought* 7 (4): 439–458.

Frey, Bruno S., and Reto Jegen. 2001. Motivation Crowding Theory. *Journal of Economic Surveys* 15 (5): 589–611.

Frederick, Shane, Loewenstein, George, and O'Donoghue, Ted. 2004. Time Discounting and Time Preference: A Critical Review. In *Advances in Behavioral Economics*, eds. Camerer, Colin F, Loewenstein, George, and Rabin, Matthew, 162–222. Princeton: Princeton University Press.

Gneezy, Uri, and Aldo Rusticchini. 2000a. Pay Enough or Don't Pay at All. *Quarterly Journal of Economics* 115 (3): 791–810.

———. 2000b. A Fine Is a Price. *Journal of Legal Studies XXIX*: 1–17.

Gordon, Myron J., and Jeffrey S. Rosenthal. 2003. Capitalism's Growth Imperative. *Cambridge Journal of Economics* 27: 25–48.

Gowdy, John. 2008. Behavioral Economics and Climate Change Policy. *Journal of Economic Behavior & Organization* 68 (3–4): 632–644.

Gowdy, John, and Jon D. Erickson. 2005. The Approach of Ecological Economics. *Cambridge Journal of Economics* 29: 207–222.

Hartley, J. E. (1997). *The Representative Agent in Macroeconomics*. London: Routledge.

Heise, Arne. 2016. 'Why Has Economics Turned Out This Way?' A Socio-Economic Note on the Explanation of Monism in Economics. *Journal of Philosophical Economics* X (1): 81–101.

Inglehart, Ronald, and Christian Welzel. 2005. *Modernization, Cultural Change, and Democracy—The Human Development Sequence*. Cambridge: Cambridge University Press.

Jackson, Tim. 2009. *Prosperity Without Growth: Economics for a Finite Planet*. Routledge.

Johnson-Laird, P.N. 1983. *Mental Models: Towards a Cognitive Science of Language, Inference, and Consciousness*. Cambridge: Cambridge University Press.

Johnson-Laird, Philip Nicholas, and Ruth M.J. Byrne. 1991. *Deduction*. Lawrence Erlbaum Associates, Inc.

Kaldor, Nicholas. 1957. A Model of Economic Growth. *Economic Journal* 67 (268): 591–624.

Kirman, A. P. 1992. Whom or what does the representative individual represent? *Journal of Economic Perspectives* 6 (2): 117–136.

Lavoie, Marc. 2006. *Introduction to Post-Keynesian Economics*. London: Palgrave Macmillan.

Levy, David L. and Andre Spicer. 2013. Contested Imaginaries and the Cultural Political Economy of Climate Change. *Organization* 20 (5): 659–678.

Livni, Joseph. 2019. Investigation of Collapse of Complex Socio-Political Systems Using Classical Stability Theory. *Physica A: Statistical Mechanics and Its Applications* 524: 553–562.

Maslow, Abraham. 1954. *Motivation and Personality*. New York: Harper & Row.

Max-Neef, Manfred. 1991. *Human Scale Development*. New York and London: The Apex Press.

Miller, Thaddeus R., Arnim Wiek, Daniel Sarewitz, John Robinson, Lennart Olsson, David Kriebel, and Derk Loorbach. 2014. The Future of Sustainability Science: A Solutions-Oriented Research Agenda. *Sustainability Science* 9: 239–246.

Mullainathan, Sendhil, and Richard H. Thaler. 2000. Behavioral Economics. No. w7948. National Bureau of Economic Research.

Negru, Ioana. 2009. Reflections on Pluralism in Economics. *International Journal of Pluralism and Economics Education* 1 (1/2): 7–21.

Negru, Ioana, and Anca Negru. 2017. Modes of Pluralism: Critical Commentary on Roundtable Dialogue on Pluralism. *International Journal of Pluralism and Economics Education* 8 (2): 193–209.

Nelson, Julie A. 2008. Economists, Value Judgments, and Climate Change. *A View from Feminist Economics. Ecological Economics* 65 (3): 441–447.

Nelson, Richard R., and Sidney G. Winter. 1982. *An Evolutionary Theory of Economic Change*. Cambridge, Mass.: The Belknap Press of Harvard Univ. Press.

———. 2007. A Review of the Stern Review on the Economics of Climate Change. *Journal of Economic Literature* 45 (3): 686–702.

——— 2013. The Climate casino: Risk, uncertainty, and economics for a warming world. New Haven, Conn: Yale University Press.

———. 2019. Climate Change. The Ultimate Challenge for Economics. *American Economic Review* 109 (6): 1991–2014.

Nordhaus, William D. 1992. An Optimal Transition Path for Slowing Climate Change. *Science* 20: 1315–1319.

Paavola, Jouni, and W. Neil Adger. 2005. Institutional Ecological Economics. *Ecological Economics* 53 (3): 353–368.

Parker, A.M., W.B. de Bruin, and B. Fischhoff. 2007. Maximizing vs. Satisficing: Decision-making Styles, Competence, and Outcomes. *Judgment and Decision Making* 2: 342–350.

Pindyck, Robert S. 2013. Climate Change Policy: What Do the Models Tell Us? *Journal of Economic Literature* 51 (3): 860–872.

———. 2017. The Use and Misuse of Models for Climate Policy. *Review of Environmental Economics and Policy* 11 (1): 100–114.

———. 2019. The Social Cost of Carbon Revisited. *Journal of Environmental Economics and Management* 94: 140–160.

Raskin, P., T. Banuri, G. Gallopin, P. Gutman, A. Hammond, R. Kates, and R. Swart. 2002. *Great Transition: The Promise and Lure of the Times Ahead.* Boston: Stockholm Environmental Institute.

Reardon, Jack, ed. 2015. Roundtable Dialogue on Pluralism. *International Journal of Pluralism and Economics Education* 6 (3): 272–308.

Richters, Oliver, and Andreas Siemoneit. 2017. Consistency and Stability Analysis of Models of a Monetary Growth Imperative. *Ecological Economics* 136: 114–125.

———. 2019. Growth Imperatives. Substantiating a Contested Concept. *Structural Change and Economic Dynamic* 851: 126–137.

Robert S., and Daniel L. Rubinfeld. 2018. *Microeconomics.* München: Pearson Studium.

Roos, Michael. 2016. *Modeling Radical Uncertainty and Anticipating Uncertain Change with Models, Forum for Social Economics.* https://doi.org/10.1080/07360932.2016.1229631

———. 2018. More foresight, less forecasting. *Discussion paper.* 1–22.

van Rooij, Maarten C.J., Annamaria Lusardi, and Rob J.M. Alessie. 2012. Financial Literacy, Retirement Planning and Household Wealth. *Economic Journal* 122 (560): 449–478.

Samuelson, Paul Anthony, and William D. Nordhaus. 2009. *Economics.* 18th edn. Boston, Mass.: McGraw-Hill/Irwin.

Sandel, Michael J. 2012. *What Money Can't Buy: The Moral Limits of Markets.* New York: Farrar, Straus and Giroux.

Satz, Debra. 2010. *Why Some Things Should Not Be for Sale—The Moral Limits of Markets.* Oxford: Oxford University Press.

Schumpeter, Joseph A. 1954. *History of Economic Analysis.* London: Allen & Unwin Ltd.

Schwartz, Shalom H. 1992. Universals in the Content and Structure of Values: Theoretical Advances and Empirical Tests in 20 Countries. *Advances in Experimental Social Psychology* 25 (1): 1–65.

———. 1994. Are There Universal Aspects in the Structure and Contents of Human Values? *Journal of Social Issues* 50 (4): 19–45.

———. 2012. An Overview of the Schwartz Theory of Basic Values. *Online Readings in Psychology and Culture* 2 (1). https://doi.org/10.9707/2307-0919.1116.

Sent, Esther-Mirjam. 2004. Behavioral Economics: How Psychology Made Its (Limited) Way Back into Economics. *History of Political Economy* 36 (4): 735–760.

Skidelsky, Edward, and Robert Skidelsky. 2012. How Much Is Enough? Money and the Good Life. Penguin UK.

Steckel, Jan, and Jörg Peters. 2019. Keine Moralpredigt, bitte! Die Zeit Online, September 5, 2019. https://www.zeit.de/wirtschaft/2019-09/klimaschutz-oekologischer-fussabdruck-industrie-co2-emissionen-verzicht-gewissen-verantwortung

Steigleder, Klaus. 2018. Climate Economic and Future Generations. In *Towards the Ethics of a Green Future. The Theory and Practice of Human Rights for Future People*, ed. M. Düwell, G. Boss, and N. van Steenbergen, 131–153. London and New York: Routledge.

Stern, Nicholas. 2007. *The Economics of Climate Change—The Stern Review*. Cambridge: Cambridge University Press.

Tainter, Joseph. 1988. *The Collapse of Complex Societies*. Cambridge University Press.

Tanner, Thomas, and Jeremy Allouche. 2011. Towards a New Political Economy of Climate Change and Development. *IDS Bulletin* 42 (3): 1–14.

Thaler, Richard, and Cass Sunstein. 2003. Libertarian[sic!] Paternalism. *American Economic Review* 93: 175–179.

van den Bergh, Jeroen C. J. M. 2007. Evolutionary thinking in environmental economics. *Journal of Evolutionary Economics Volume* 17 (5): 521–549.

Vatn, Arild. 2009. Cooperative Behavior and Institutions. *Journal of Socio-Economics* 38: 188–196.

Weitzman, Martin L. 2009. On modelling and interpreting the economics of catastrophic climate change. *Review of Economics and Statistics* (91): 1–19.

Wren-Lewis, Simon. 2007. Are There Dangers in the Microfoundations Consensus? In *Is There a New Consensus in Macroeconomics?* ed. P. Arestis. Palgrave Macmillan.

———. 2011. Internal Consistency, Nominal Inertia and the Microfoundations of Macroeconomics. *Journal of Economic Methodology* 18 (2): 129–146.

# CHAPTER 5

# Why We Do Not Have More Pluralism

**Abstract** This chapter explains why there is little pluralism in economics. Mainstream economists believe that journal rankings and citation measures are strong indicators of research quality and that science is like a free market of ideas. Roos and Hoffart argue that these beliefs are wrong and lead to a systematic bias against non-mainstream economics. Neoclassical economics desires to emulate the 'hard science' of physics leading to a very restricted methodology. Since most mainstream economists have little training and interest in theory of science, they avoid methodological discussions and are unable to engage in a fruitful exchange with non-mainstream researchers. Such exchange is a main element of interested pluralism. The authors argue that strongly standardised teaching based on textbooks is a major hindrance for pluralism.

**Keywords** Pluralism • Economic curriculum • Superiority • Scientific market • Theory of science

## 5.1  Introduction

Current economics as a whole is a monist discipline (King 2013) characterised by a distinction between an *orthodoxy* and a *heterodoxy*. The wording of *orthodox* vs. *heterodox* used in economics is stronger than the

M. Roos, F. M. Hoffart, *Climate Economics*, Palgrave Studies in Sustainability, Environment and Macroeconomics, https://doi.org/10.1007/978-3-030-48423-1_5

distinction between *conventional* vs. *unconventional* or *mainstream* vs. *sidestream* approaches common in other social sciences (Davis 2007). *Orthodox* implies "being generally accepted" or "being right", whereas *heterodox* means "being at variance with established or accepted beliefs" with the connotation of "being wrong". If mainstream economists use the label heterodox for non-mainstream research, they often insinuate that it is bad science or maybe not even science at all.

While it is probably true that some heterodox economists prefer to criticise the mainstream instead of engaging into a constructive dialogue[1]—which we called *disinterested pluralism* in the previous chapter— there are also exclusion mechanisms in the mainstream. These exclusion mechanisms make it hard for non-mainstream research to gain recognition or become mainstream, even if this is the aim of the respective researchers.

In this chapter, we explain why there is a dominance of neoclassical thinking in economics. We argue that there are general tendencies towards a paradigmatic concentration present in many disciplines as they are currently organised. These monopolisation tendencies contradict the concept of a *free market of ideas* that many economists maintain (Stigler 1988). But in addition to mechanisms that stabilise a dominant paradigm in any discipline, there are specific particularities in economics making economics arguably less pluralist than other social sciences. The first particularity is the neoclassicals' emulation of what they perceive the "scientific approach" inspired by classical physics and the resulting feeling of superiority with regard to other social sciences. A second particularity is the widespread disdain of methodological discussions which coincides with a low level of training in formal methodological reasoning and philosophy of science. Finally, there is a strong standardisation of academic teaching at all levels, including PhD programs. Addressing these particularities would be key to enhance pluralism in economics, which we consider important.

## 5.2   No Free Market of Ideas

For economists, it is natural to think of science as a market of ideas, in which better ideas are more successful than worse ones (Klein 2005). According to Stigler (1988, pp. 84–85)

---

[1] See   http://www.philosophersbeard.org/2010/10/why-is-heterodox-economics-joke. html for a harsh characterisation of the style of some heterodox economists.

Scientific research is a market process, differing vastly in form but little in substance, from the comparable activities of grocers or manufacturers of computers. Individual scholars distribute themselves by the action of self-interest. ... The better the research work, the more prestigious the journal in which it will appear. The superior researcher is hired by the better university, promoted at a rapid rate, favored by the National Science Foundation and the private foundations, given a lighter teaching load. ... There are so many good universities ..., so many scientific journals, that each major discipline becomes a competitive industry. No one model of orthodoxy can be imposed on a science.

The analogy between the scientific system and a competitive market is flawed for several reasons and an example of misled neoclassical reasoning. The key fallacy consists in the wrong assumption that the supply of scientific ideas is separated from their demand as it is the case in a market. The first recipients of ideas produced by scientists are other scientists. Only at later stages, after the scientific ideas have been transformed into scientific knowledge in form of publications, other actors from outside of academia have a demand for this knowledge. But whether ideas, theories or evidence count as scientific knowledge is determined inside the academic system. This is very different in markets, because producers can define their products independent of other producers. A manufacturer of computers does not need recognition from other manufacturers of computers that the product offered actually *is* a computer. Other manufacturers may have better products, but it is upon the customers to make this judgement. A second fallacy is to assume that incumbents in science are challenged by new competitors. In order to be allowed to judge whether some work is scientific, one must be acknowledged as a competent scientist by the scientific community. The recognition as a scientist is achieved by obtaining a PhD and being visible in the community by publications and an affiliation at an academic institution. Academic gatekeepers in the form of PhD supervisors, journal editors and referees as well as members of hiring committees strongly regulate the entry to academia and related success.

Maybe more than people working in other professions, scientists are driven by a particular mix of extrinsic and intrinsic motivation. Scientists care about income, professional reputation and other benefits provided by their positions. In addition to this extrinsic motivation, however, many scientists are intrinsically motivated to do science. They have an intrinsic interest in their objects of study and derive utility from new insights. Motivations interact with institutions, because institutions set incentives. Furthermore, institutions give guidance on what is the "right thing" to do (Vatn 2009). Depending on the institutional context, it might be appropriate either to follow an individual rationality or a cooperative social rationality. Applied to science, this means that the belief how research should be done and what defines good research is a matter of institutions that provide incentives and thus shape the beliefs of the scientific community.

In economics, researchers are expected to publish scientific papers in peer-reviewed journals. Other formats such as books, newspaper articles, conference presentations or internet blogs have little value as scientific output. The productivity of a researcher is assessed in terms of the quantity and quality of her journal publications. There is the strong conviction in economics that the quality of an individual work of research can be measured by the quality of the journal in which it is published. The quality of journals, in turn, is reflected in journal rankings or ratings, which are typically based on journal impact factors calculated mostly from the citations a journal receives. There is a considerable number of journal rankings in economics and also a lot of research on the best way to construct these rankings. Bornmann et al. (2017) conducted a meta-study over 22 rankings of general economics journals. Some journal rankings are also (partly) based on expert opinion like the journal list of the Tinbergen Institute or the survey-based list in Bräuninger and Haucap (2003). It is a common practice to attach labels to the journals such as *A+, A, B, C* and/ or *top journal, excellent journal, good journal, minor journal* (e.g. Ritzberger 2008).

Not only journals are ranked in economics but also economics departments at universities (e.g. McPherson et al. 2013), research institutes and individual researchers. The RePEc (Research Papers in Economics) project[2] is an effort by volunteers to build a decentralised bibliographic database which is then used to compute various rankings of individuals and organisations. Lee and Cronin (2010) propose a ranking of heterodox journals, arguing that this is necessary because mainstream economics and heterodox economics are distinct bodies of knowledge that cannot be compared directly. According to their ranking, the best heterodox journals are: *Cambridge Journal of Economics, Journal of Economic Issues, Journal*

---

[2] http://repec.org/.

*of Post Keynesian Economics, Review of Radical Political Economics,* and *Economy and Society.* In the Ritzberger (2008) list, only the first two appear under the top 261 journal and get the label "C: minor journals". In the Bornmann et al. (2017) list, the Cambridge Journal is ranked 64, the Journal of Economic Issues is on rank 183 and the Journal of Post Keynesian Economics receives rank 199. If heterodox research is not acknowledged to be distinct from mainstream research and evaluated by mainstream rankings, the conclusion will be that it is of inferior quality.

Professional success in economics depends strongly on these rankings (Frey and Rost 2010; Frey 2003). The perceived research quality is a crucial determinant for hiring decisions in economic departments (McPherson et al. 2013; Bornmann et al. 2017), salaries (Gibson et al. 2014), for funding from research funding institutions (Cole et al. 1981) and for professional prestige. The shared conviction that journal rankings and citations are a reliable indicator of research quality is an informal institution, such as the general willingness to use these indicators for career-relevant decisions, for instance in hiring committees. The way in which the publication of research in peer-reviewed journals is organised is a formal institution. These institutions set incentives for researchers and shape their beliefs how proper science should be.

The strong belief of economists in rankings would not be a problem, if they were an adequate reflection of research quality. In this case, the best researchers would publish in the best journals and work at the best institutions. In fact, the U.S. academic system is especially hierarchically organised as if this were the case: researchers working at the top institutions publish in the top journals and vice versa (Lee 2006). Gloetzl and Aigner (2017) show that economics is a highly concentrated discipline in terms of publications, institutions, citations, authors and regions. For example, the "Top Five" journals (*American Economic Review, Quarterly Journal of Economics, Journal of Political Economy, Econometrica* and *Review of Economic Studies*) out of 675 journals account for 27.7% of all citations between 1956 and 2016. Around half of all authors are affiliated with North American institutions and about three quarters of all citations are received by North American authors. The top 20 academic institutions out of about 4300 academic institutions, of which 18 are located in the U.S., account for about 15% of all articles and more than 42% of all citations. Note that two of the top-5 journals are based at the two leading U.S. universities in terms of the number of citations and published articles:

the *Quarterly Journal of Economics* is published by Harvard University and the *Journal of Political Economy* is published by the University of Chicago.

There are good reasons to doubt that the assumed close correspondence between research quality, institutions, journal publications and citations is really valid. In particular, it can be questioned that the ranking of a journal is a good indicator of quality of the individual papers published there. It is neither true that every article in a top journal and every researcher at a top institution are excellent nor that every article in lower-ranked journals and every researcher at a lower-ranked institution are of minor quality.

Seglen (1992) shows that the impact factor of a journal is heavily influenced by a minority of articles. With data from three biochemistry journals, he shows that 15% of the articles account for 50% of all citations and that the most cited 50% of the articles receive 90% of the journals' citations. Oswald (2007) analyses the impact in terms of received citations of a selection of articles published in 1981 in first-tier journals (American Economic Review and Econometrica), second-tier journals (Journal of Public Economics and Economic Journal) and third-tier journals (Journal of Industrial Economics and Oxford Bulletin of Economics and Statistics). He finds that the median number of citations after 25 years of all articles in the respective issues of the two first-tier journals is 22, but the four least-cited articles in both journals received a sum of only 6 (AER) and 5 (Econometrica) citations together. He concludes that ex post, these articles did not have a major scientific impact and should not have been published in the top journals. The best-cited papers in the lower-ranked journals received between 50 and 199 citations and hence had a much bigger impact, which might have justified the publication in a higher-ranked journal. Laband and Tollison (2003) speak of "dry holes" in economic research, that is, papers that receive no citations five years after publication. They show that in 1996 about 26% of all papers in 91 economics journals did not receive a single citation. They argue that the peer review system failed to detect these barren papers despite considerable screening investment. Some authors even argue that peer review is not only ineffective but also harmful to science and should be abolished (Osterloh and Frey 2015). There are at least three reasons, why peer review can fail to evaluate the quality adequately: (1) mistakes by reviewers and editors and chance, (2) systematic biases, and (3) strategic and selfish behaviour.

Benda and Engels (2011) argue that peer review has strong similarities to judgmental forecasting in other domains and is hence always subject to error and randomness (also Neff and Olden 2006). Cole et al. (1981) conducted an experiment in which grant applications to the U.S. funding institution NSF were reviewed again by different reviewers. The correlation between the mean ratings of the proposals in the original review and the experimental review are moderately high (between 0.6 and 0.66). More importantly, they find a significant percentage of reversals of the initial funding decision (between 24% and 30%). In contrast to chemical dynamics and solid-state physics, the percentage of reversals in economics was highest in the second and the fourth quintile of the quality rating (45% and 42%, respectively) and not in the middle quintile, where one might expect the quality differences between funded and non-funded proposals to be lowest. Reviewers' disagreement concerning the quality of manuscripts and research proposals is frequent. Cole et al. (1981, p. 885) conclude that "the fate of a particular grant application is roughly half determined by the characteristics of the proposal and the principal investigator, and about half by apparent random elements which might be characterized as the 'luck of the reviewer draw'".

One can argue that random elements in peer-reviewing are a general problem in science, but not a particular one for non-mainstream approaches, because they affect mainstream research, too. Yet, there is evidence that systematic biases in the evaluation of research quality exist. In a well-known paper, Merton (1968) demonstrated a *Matthew Effect* in science, according to which "eminent scientists get disproportionately great credit for their contributions to science while relatively unknown scientists tend to get disproportionately little credit for comparable contributions" (Merton 1968, p. 57). Other papers empirically confirm the existence of the Matthew effect (e.g. Bonitz et al. 1999; Larivière and Gingras 2009; Medoff 2006; Birkmaier and Wohlrabe 2014; Tol 2009, 2013). The Matthew effect can mean that a well-known author receives relatively more citations for an article than an unknown researcher for a paper of similar quality. More importantly, papers have a higher chance of being published in top journals, if their authors are affiliated with prestigious research institutions than if they originate from unknown ones. Peters and Ceci (1982) conducted an experiment in which they resubmitted articles that were already published in high-ranked psychology journals to the same journals 18 to 32 months later, but with fictitious names of authors and institutions. They found that only three out of 38 editors and reviewers detected the resubmissions. Out of nine re-reviewed articles, only one was accepted. 16 out of 18 referees recommended against publication, mostly because of alleged methodological flaws. Peters and Ceci (1982) conclude that the only plausible reason for their findings is that both reviewers and editors are biased in favour of authors and institutions with high reputation and against unknown ones.

Larivière and Gingras (2009) report the findings of a natural experiment concerning citations of articles in many different scientific disciplines. They identify 4532 pairs of duplicate papers that appeared in journals with different impact factors but had the exact same title, the same first author and the same number of cited references. Papers published in higher ranked journals on average receive a significantly higher number of citations than their lower ranked duplicates. Medoff (2006) finds that economics papers published by authors from Harvard University and the University of Chicago are cited disproportionately more often than papers of comparable quality, but whose authors come from other U.S. (elite) universities. Klamer and van Dalen (2002) suggest that one reason for the Matthew effect is scarce attention. It is impossible for researchers to pay attention to every piece of work published somewhere. Focusing one's attention mostly on high-ranked journals and the activities of well-known scientists at prestigious institutions can hence be a reasonable way to deal with information overload.

Finally, it cannot be ruled out that scientists in their roles as authors, reviewers and editors have selfish interests that conflict with the interests of the scientific community as a whole. It is reasonable that many care about their own careers and the careers of other researchers with whom they have a personal relation. Given the time constraint, every scientist is faced with the question on how much time to allocate to the own research and to reviewing the work of others. Since reviewing under the current system has lower returns for one's own career than publishing, there is an incentive to minimise reviewing time and effort, which might affect the quality of the reviews. Reviewers with moderate and low reputation might fear that their reputation is harmed if they recommend very innovative research for publication. Such research has a higher risk of being ignored or criticised by others such that the reviewers must be willing to make a subjective recommendation about the potential impact of the research. Even established scientists with a high reputation may be reluctant to accept novel research that is different from or in contrast to their own work. They might either fear losing reputation if their own work is criticised or proven to be flawed or they might have personal ontological and epistemological convictions how good science should be.

Researchers could also be tempted to ignore or suppress research by perceived competitors. The scientific system is often compared to a *winner-takes-it-all market* (Klamer and van Dalen 2002), creating strong incentives for behaving competitively. Not without reason, many journals and funding institutions require reviewers to declare if they have a conflict of interest, which is hard to verify. Researchers and institutions also have an incentive to promote their own PhD students or colleagues, because it increases the reputation of their own institution. Klein (2005) provides evidence that 90% of faculty members of the global top 35 economics departments received their PhDs from exactly those 35 departments. Hodgson and Rothman (1999) and Klamer and van Dalen (2002) discuss the empirical and theoretical importance of close networks among researcher at top economic departments in their roles as authors, editors and reviewers.

Recently, many researchers applied agent-based models in order to analyse the outcomes of the peer review process under different assumptions about the motivations of the scientists and the institutions (Sobkowicz 2015; Thurner and Hanel 2011; Kovanis et al. 2016; Allesina 2012; Squazzoni and Gandelli 2012; Squazzoni and Gandelli 2013). A general message of these papers is that even small numbers of "imperfect" agents that are lazy, incompetent, selfish or reciprocal to peers can distort the system significantly. Thurner and Hanel (2011) demonstrate with shares of one third each of rational, random and correct reviewers the peer review system does not perform better than choosing papers randomly. In their model, rational reviewers do not accept work that is better than their own and random reviewers are either lazy or incompetent and hence give random recommendations. Another important result of this research is a strong bias against innovation (Sobkowicz 2015). The suppression of innovative research by peer review is also confirmed empirically (Armstrong 1997; Fölster 1995; Gans and Shepherd 1994; Benda and Engels 2011).

The strong reliance of economics on rankings, peer review and citations as measures of scientific quality has significant implications for heterodox approaches. First, there is a deterrence effect. If non-mainstream research is generally perceived to be inferior instead of just different, because it is published in lower-ranked journals, researchers' intrinsic motivation to pursue this kind of research can be harmed. Frey (2003, p. 206) compares publishing in refereed journals with prostitution:

The system of journal editing existing in our field at the present time virtually forces academics to become prostitutes: they sell themselves for money (and a good living). Unlike prostitutes who sell their bodies for money ..., academics sell their soul to conform to the will of others, the referees and editors, in order to gain one advantage, namely publication. Most persons refusing to prostitute themselves and to follow the demands of the system are not academics: they cannot enter, or have to leave, academia because they fail to publish. Their integrity survives, but the persons disappear as academics.

Osterloh and Frey (2015) argue that extrinsic rewards can crowd out the intrinsic *taste for science*, which is characterised by a high importance of peer recognition and autonomy. If peer recognition depends on rankings, scientists have an incentive to game the system. By external monetary and reputational rewards based on rankings, scholars lose their autonomy, which is reported to be an important factor of intrinsic motivation and creative research (Amabile et al. 1996; Amabile 1998). Ashforth (2005, p. 400) provocatively labels many researchers as "vanilla pudding: comfort food, yes, but ultimately a little bland, with a lot of empty calories". What he means is that young researchers are socialised to do uncreative, trivial and often irrelevant research, because it would be very risky to do unconventional research before one has tenure. The will and the ability to work more creatively are often lost after the long process of getting tenure, which requires publishing many conventional papers.

The second implication of how economics is currently organised is that heterodox economists have lower chances to get good positions in research-oriented organisations, to receive research funding and to attract ambitious and smart PhD students. Therefore, their material preconditions for productive research are worse than those of mainstream economists.

### 5.3    Emulation of Physics and Feeling of Superiority

For mainstream economists, it is important to be regarded as true scientists. Referring to the thriving U.S. educational and job market, Mayer (1993, p. 1) writes about the general opinion of economists:

Not only does the market signal its approval, but this approval seems justified by the increasing scientific nature of economics. No longer can economists be looked down on as would-be scientists, who cannot even speak what is widely regarded as the language of science, mathematics. Mathematical economists can now converse in that tongue as fluently as physicists. Powerful theorems are being discovered all the time. ... Economics no longer requires the unique insights of great thinkers to make any progress but can proceed step-by-step as a normal science, just like the natural sciences do.

This opinion expresses a great concern about status in economics, which is also reflected in the importance of rankings discussion in the previous subsection. This preoccupation with status does not only apply to the comparisons of economists among themselves but also to comparisons with other disciplines. Economists like to believe that they have a special status among the social sciences:

Not only is science the paradigm of specialized and objective knowledge, but scientists are also accorded high status because of the material achievements that science has fostered. But not all sciences. The social sciences, particular sociology, have low status. Hence, while economists can hardly deny that economics is a social science, they have a strong incentive to differentiate economics from other social sciences, and to claim a kinship to the hard sciences, in spirits, in mental toughness, in technique, and even in achievement. Generally conceded to be the most advanced of the social sciences, economics would love to drop the bar sinister of a social science and take is rightful place as a member of the true scientific aristocracy. (Mayer 1993, p. 15)

Fourcade et al. (2015) describe that economists have a feeling of superiority with regard to the other social sciences and explain where it comes from:

[Economics] is characterised by far-reaching scientific claims linked to the use of formal methods; the tight management of the discipline from top down; high market demand for services, particular from powerful and wealthy parties; and high compensation. This position of social superiority also breeds self-confidence, allowing the discipline to retain its relative epistemological insularity over time and fueling a natural inclination towards a sense of entitlement. (Fourcade et al. 2015, p. 91)

Economists derive their self-perception and self-confidence to a large extent from the use of formal, mathematical methods, which is much less frequent in other social sciences. Colander (2005) reports the results of a survey study among graduate students in economics at seven top U.S departments. In this survey, 77% of the students agreed with the statement that "economics is the most scientific of the social sciences" and 89% believed that "neoclassical economics is relevant". Asked about important factors for success in economics, large shares of the respondents agree with "Excellence in mathematics" (82%) and "Being interested in, and good at, empirical research" (82%), but only one-third of the graduate students believe that "a thorough knowledge of the economy" is a success factor.

These particular attitudes in mainstream economics are deeply interwoven with the neoclassical approach. The neoclassical methodology has its origins in the physics of the late nineteenth century (Mirowski 1984, 1989, 1991). Neoclassical economists who had received training as engineers and in physics, such as Léon Walras, William Stanley Jevons, Francis Ysidro Edgeworth and Irving Fisher, were protagonists of the *marginalist revolution* starting in the 1870s. Mirowski (1989) shows that these economists did not only adapt the methodology in the physics of their time but even copied the equations and replaced the physical variable by economic ones. Irving Fisher used the proto-energetics model from thermodynamics to develop the now standard neoclassical consumer choice theory. Mirowski (1989, p. 224) reproduces a list, in which Fisher translates variables from mechanism into economic variables. For instance, *utility* corresponds to *energy* and *marginal utility* corresponds to *force*. Importantly, the neoclassical economists also took over the concept of *equilibrium* from physics. The first attempts to formalise economics were viewed with great scepticism by other economists and remained a niche until the mid-1920s (Mirowski 1991). Starting in the mid-1920s, a second wave of trained scientists and engineers turned to economics and updated the neoclassical research program with more recent methods they knew from physics and mathematics. Leading figures were Ragnar Frisch, Tjalling Koopmans, Jan Tinbergen, Maurice Allais and Kenneth Arrow. Optimisation methods, like the Lagrange function, the Hamiltonian function and the calculus of variations, which are now standard tools of mainstream economic analysis, were imported from physics in this second wave of formalisation. Other formalisms in economics with these origins are the use of stochastics, phase space and linear algebra.

The desire to make economics more scientific and to use the powerful mathematical tools common in physics guided economists to specific problems to which these tools could be applied and shaped their thinking about the economy. It is hence not surprising that neoclassical economics has a rather mechanistic view of economic agents and focuses on those aspects of economic life that are easy to formalise and ignores the messier ones. The neoclassical insistence that economics is the science of how to allocate scarce means optimally to competing ends can be seen as a consequence of availability of optimisation methods.

## 5.4 Avoidance of Theory of Science

Mainstream economists often argue that it is false to diagnose a lack of pluralism in economics and bring forward two arguments for their claim. The first argument for the alleged pluralism is the variety of topics that are covered by economists, for example, from the economics of aging over defence economics to wine economics. When confronted with the argument that a variety of topics is not the same as pluralism of approaches, they retort that economists also have a diverse tool set, such as game theory, DSGE models, a host of econometric models and experiments in the lab and in the field. If a critic counters again by saying that true pluralism does not happen at the level of *methods*, but implies significant differences in methodology, epistemology and, most importantly, ontology, the discussion is likely to end leaving the mainstream economist in incomprehension.

Mainstream economists know very little about theory of science and often actively avoid getting involved in methodological debates (Lawson 1994; Heise 2016; Drakopoulos 2016). Some economists even express an open hostility towards methodology. In 1992, an interchange between the Cambridge theorist Frank Hahn and the methodologist Roger Backhouse took place via the Royal Economic Society Newsletter. Hahn (1992a) had triggered this debate with critical reflections on methodology, prompting Backhouse (1992) to ask "Should we ignore methodology?" to which Hahn (1992b) replied: "Answer to Backhouse: Yes". In his "Reflections" Hahn (1992a) advises young economists to "avoid discussion of 'mathematics in economics' like the plague" and to "give no thought at all to methodology". Methodology should be left to philosophers and would not help practising economics. There is the pejorative saying in economics that "those who can do while those who cannot discuss methodology" (see Lawson 1994). Since it is often non-mainstream economists who are concerned with methodological issues and use methodological arguments to criticise the mainstream, they are vilified as bad scientists not only because of their different approaches but also due to their interest in methodology. Many mainstream economists perceive methodology as a prescriptive set of rules for good science, which corresponds to a traditional understanding of methodology based on logical positivism (Dow 2009). They are not aware of the developments in philosophy of science that challenge logical positivism and call for a fundamental rethinking of how science works.

After the global financial and economic crisis 2007–2008, Colander et al. (2009) published a scathing critique of mainstream macroeconomics, diagnosing a "systemic failure of academic economics". The main point of the critique is that mainstream macroeconomists constructed models with a weak methodological basis and poor empirical performance. Since the available DSGE models could not even *in principle* generate crises of the type that happened in reality, they are regarded as a scientific failure calling for a major reorientation of macroeconomics, which required a fundamental methodological debate. However, this fundamental methodological debate did not take place, as the contributions in the special issue of the Oxford Review of Economic Policy on the "rebuilding macroeconomic theory project" (Vines and Wills 2018) show. The debate remains at the level of the *method*, but neither discusses methodology nor deeper ontological problems. Tellingly, even the discussion on how to improve the heavily criticised DSGE models is reluctant and accompanied by introductory excuses. Reis (2018) begins his article titled "Is something really wrong with macroeconomics?" as follows:

> I accepted the invitation to write this essay and take part in ths debate with great reluctance. The company is distinguished and the purpose is important. I expect the effort and arguments to be intellectually serious. At the same time, I call myself an economist and I have achieved a modest standing in this profession on account of (I hope) my ability to make some progress thinking about and studying the economy. I have not expertise in studying economists [in italics in the original]. (Reis 2018, p. 132)

Flirting with doing something inappropriate, Blanchard, who had written on the state of macroeconomics earlier (e.g. Blanchard 2009, 2016), introduces his essay "On the future of macroeconomic models" with the statement:

One of the best pieces of advice Rudi Dornbusch gave me was: never talk about methodology; just do it. Yet, I shall make him unhappy, and take the plunge. ... now, with this article, [I write] my final piece (I promise) on this topic. (Blanchard 2018, p. 43)

Several authors attempt to explain the mainstream's hostility towards methodology. Arnsperger and Varoufakis (2006) argue that the lack of methodological debate is an (unintentional) act of self-protection. If mainstream economists engaged in methodological discussions, they would have to admit that many of their key assumptions are hard to defend empirically. Furthermore, methodological discussions would undermine their discursive power that earns them influence and resources. Drakopoulos (2016) explains the disdain for methodology with the physics envy of mainstream economists. Methodological discussions unavoidably would centre on the question whether approaches imported from thermodynamics are really appropriate for the study of social systems. Since the likely answer would be "no", economics' status of a hard science like physics would be threatened. Lawson (1994) elaborates the argument that the mainstream's endorsement of positivism as the philosophical basis of its research program explains the denigration of methodology. O'Sullivan (2019) relates the avoidance of discussions of normative questions in economics to the mathematisation of economics and orientation towards physics. The break between ethics and economics, he describes as the reason for "a serious erosion of any sense of personal ethical responsibility among professional economists" (O'Sullivan 2019, p. 48), also leads to the avoidance of theory of science questions.

## 5.5   Strong Standardisation in Teaching

Mainstream economics' monism is also reflected in academic teaching, which contributes to its perpetuation. There is widespread dissatisfaction among students of economics about what they are taught and especially what they are not taught. In June 2000, a group of students of the elite *Grandes Écoles* in France started the so-called *post-autistic economics* movement with a petition published on the web. They protested against (Fullbrook 2003, p. 1):

- Lack of realism in the teaching of economics
- Economics as an autistic science, lost in imaginary worlds due to the uncontrolled use of mathematics
- The repressive domination of neoclassical theory and
- The dogmatic teaching style and the absence of critical and reflective thought.

Among other things, the French students argued in favour of more pluralism in economic teaching and more engagement with the real world, instead of studying abstract models. What began as a protest by a small group of students in France quickly spread to other countries and turned into a mass movement that received considerable attention in the media and in politics. The French government asked the eminent economist Jean-Paul Fitoussi to lead a commission set up to investigate the complaints of the students. In its report that appeared in 2001, the commission basically supported the students' complaints and called for more debate on contemporary economic issues in university education and for much more multidisciplinarity (Fullbrook 2003, pp. 5–6).

The post-autistic economics movement, which later changed its name to *movement of students for the reform of economics teaching*, received a boost by the global financial crisis that was widely perceived as a failure of the science of economics. In 2014, the *International Student Initiative for Pluralism in Economics* published an open letter (http://www.isipe.net/open-letter) with the following opening paragraph:

> It is not only the world economy that is in crisis. The teaching of economics is in crisis too, and this crisis has consequences far beyond the university walls. What is taught shapes the minds of the next generation of policymakers, and therefore shapes the societies we live in. We, over 65 associations of economics students from over 30 different countries, believe it is time to reconsider the way economics is taught. We are dissatisfied with the dramatic narrowing of the curriculum that has taken place over the last couple of decades. This lack of intellectual diversity does not only restrain education and research. It limits our ability to contend with the multidimensional challenges of the 21st century—from financial stability, to food security and climate change. The real world should be brought back into the classroom, as well as debate and a pluralism of theories and methods. Such change will help renew the discipline and ultimately create a space in which solutions to society's problems can be generated.

The mainstream curriculum does not include history of economic thought and methodology. Consequently, students do not learn how the discipline evolved and do not obtain any awareness of methodological problems. And if they feel uneasy with the methodology they are implicitly taught, they do not have the means to understand what is wrong or the language to talk about it. Dow (2009) argues that students of economics should be taught both history of economic thought and methodology to equip them with the tools to form their own critical judgement on how to practice economics. The *CORE project* was initiated as a response to the demand from the students. CORE stands for *Curriculum Open-access Resources in Economics*. The main output of the project is the introductory undergraduate e-book course *The Economy*. CORE is meant to be a significant reform project to teach "economics as if the last three decades had happened" (Carlin 2016), that is, to relate it to real-world events, such as the financial crises of the 1990s and the 2000s, and to incorporate recent research into undergraduate teaching. The project leader, Wendy Carlin, presents CORE as a paradigm-setting text comparable to classics such as *Principles of Political Economy* by John Stuart Mill and *Principles of Economics* by Alfred Marshall. However, Mearman et al. (2018) came to the conclusion that it does not constitute a significant paradigm shift, but rather reproduces existing economics. In particular, they criticise that CORE does not exhibit pluralism and promotes 'instrumental' rather than 'critical education', meaning that the methodological foundations and the educational goals of the material are neither made explicit nor discussed.

Bäuerle (2017) offers an explanation for the particular way economics is taught that differs from the practice in other social sciences. He refers to Thomas Kuhn's description of scientific paradigms in the natural sciences (see Kuhn 2012) and applies it to economics. Like the natural sciences, but in contrast to other social sciences, economics is a *textbook science*. This means that economic teaching is almost exclusively done via textbooks, especially at the introductory level, but even at later stages of study. Often, students of economics start reading original research literature only at the PhD level. Even PhD training is very standardised with so-called core courses in the first year and advanced textbooks (Colander 2005). Classic texts are almost never read, and the textbooks present the history of the field (if they do) as a linear progression. It starts from now-outdated ideas to better theories until the current state of the art is reached, which is presented as the best available knowledge. Bäuerle (2017) calls Paul Samuelson's *Economics* the "archetype of the economic textbook literature", which, now in its 19th edition, is among "the most successful textbooks ever published in the field" (Skousen 1997, p. 137). Samuelson was well aware of the impact of his book, as expressed by his quote: "I don't care who writes a nation's laws—or crafts its advanced treaties—if I can write its economic textbooks" (after Elzinga 1992, p. 878). Samuelson's *Economics* was translated in 41 languages and served as a model for other widely used textbooks in terms of structure, topics, graphical representation and didactic exposition (Bäuerle 2017, p. 2).

According to Kuhn (2012), textbooks are a characteristic feature of *normal science* that does not question the rules of how science is practised and focuses on *solving puzzles*. Textbooks socialise students into normal science and set the mental framework for research within the paradigm. They define the accepted knowledge, the open questions worth being researched and the permissible methods. Bäuerle (2017) shows that in introductory textbooks economics is typically introduced as a special way of thinking and not as a field that covers specific subjects. This way of thinking manifests itself in a number of *economic principles* on which every economic analysis is based and that are presented often in the first chapter of introductory textbooks, for example, 10 principles in Mankiw and Taylor (2014) and Mankiw (2015), 7 principles in Frank et al. (2013), 8 principles Gwartney et al. (2015) and 12 principles in Krugman and Wells (2015). The role of these principles in economics is nicely illustrated by the metaphor of a tree:

> Both the structure of the economics discipline and the major itself can be likened to a giant tree. The major is rooted in the introductory courses, which introduce students to economic thinking and its applicability to a variety of issues. The trunk is a core set of principles, analytical methods, and quantitative skills that are widely accepted in the profession. The branches of the tree, extending in all directions, represent the array of subdisciplinary fields, ranging from monetary economics to industrial organization. These subfields reflect the main points of interest and research in economics and generate the problems to which principles and quantitative approaches can be fruitfully applied. (Siegfried et al. 1991, p. 202)

A similar metaphor expresses the special status of economics among the social sciences:

> These two characteristics of economics—a central core of theoretical and empirical knowledge, combined with opportunities to extend that knowledge to a wide variety of topics—differentiate it from the structure of other

social science disciplines. Whereas economics can be likened to a tree, other social sciences have a more hedge-like structure of separate and largely independent subfields with their own content and methodology. (Siegfried et al. 1991, p. 202)

Bäuerle (2017) makes the crucial point that the roots of the knowledge tree in economics are hidden below the surface and never shown. It is decisive for normal science that it never explicates its historical and methodogical origins, but silently assumes them in the textbooks. Thereby, it becomes difficult, if not impossible, for researchers to challenge the paradigm. According to Kuhn (2012) it is the nature of scientific revolutions to dig out those roots, cut them and to rewrite the textbooks.

As long as economics remains the textbook science, it is unlikely that pluralism will flourish. The very idea that there is a core of established knowledge, principles and methods, which can be codified in textbooks and taught in core courses, is at odds with the concept of pluralism.

## REFERENCES

Allesina, Stefano. 2012. Modeling Peer Review: An Agent-Based Approach. *Ideas in Ecology and Evolution* 5 (2): 27–35.

Amabile, T. 1998. How to kill creativity. *Harvard Business Review* 76: 76–87.

Amabile, T., R. Conti, H. Coon, J. Lazenby, M Herron (1996). Assessing the work environment for creativity. *Academy of Management Journal* 39: 1154–1184.

Armstrong, J. Scott. 1997. Peer Review for Journals: Evidence on Quality Control, Fairness, and Innovation. *Science and Engineering Ethics* 3: 63–84.

Arnsperger, Christian, and Yanis Varoufakis. 2006. What Is Neoclassical Economics?: The Three Axioms Responsible for Its Theoretical Oeuvre, Practical Irrelevance and, Thus, Discursive Power. *Panoeconomicus* 53 (1): 5–18.

Ashforth, Blake E. 2005. Becoming Vanilla Pudding: How We Undermine Our Passion for Research. *Journal of Management Inquiry* 14 (1): 400–403.

Backhouse, R. 1992. Should We Ignore Methodology? *Royal Economic Society Newsletter* 78: 4–5.

Bäuerle, Lukas. 2017. Die ökonomische Lehrbuchwissenschaft—Zum disziplinären Selbstverständnis der Volkswirtschaftslehre. *Momentum Quarterly* 6 (4): 210–289.

Benda, Wim G.G., and Tim C.E. Engels. 2011. The Predictive Validity of Peer Review: A Selective Review of the Judgmental Forecasting Qualities of Peers, and Implications for Innovation in Science. *International Journal of Forecasting* 27 (1): 166–182.

Birkmaier, Daniel, and Klaus Wohlrabe. 2014. The Matthew Effect in Economics Reconsidered. *Journal of Informetrics* 8 (4): 880–889.

Blanchard, Oliver. 2009. The state of macro. *Annual Review of Economics* 1: 209–228.

———. 2016. *Do DSGE model have a future?* available at https://piie.com/publications/policy-briefs/do-dsge-models-have-future.

———. 2018. On the future of macroeconomic models. *Oxford Review of Economic Policy* 34 (1–2): 43–54.

Bonitz, M., E. Bruckner, and Andrea Scharnhorst. 1999. The Matthew Index— Concentration Patterns and Matthew Core Journals. *Scientometrics* 44 (3): 361–378.

Bornmann, Lutz, Alexander Butz, and Klaus Wohlrabe. 2017. What Are the Top Five Journals in Economics?: A New Meta-Ranking. *Applied Economics* 50 (6): 659–675.

Bräuninger, Michael, and Justus Haucap. 2003. Reputation and Relevance of Economics Journals. *Kyklos* 56 (2): 175–197.

Carlin, Wendy. 2016. Teaching Economics Using the CORE Resources. *Paper presented at "Implementing and teaching CORE" conference.* Sheffield University, March 11.

Colander, David. 2005. The making of an economist redux. *Journal of Economic Perspectives* 19 (1): 175–198.

Colander, David, Michael Goldberg, Armin Haas, Katarina Juselius, Alan Kirman, Thomas Lux, and Brigitte Sloth. 2009. The Financial Crisis and the Systemic Failure of the Economics Profession. *Critical Review* 21 (2–3): 249–267.

Cole, Stephen, Jonathan R. Cole, and Gary A. Simon. 1981. Chance and Consensus in Peer Review. *Science* 214 (4523): 881–886.

Davis, J.B. 2007. The Turn in Recent Economics and Return of Orthodoxy. *Cambridge Journal of Economics* 32 (3): 349–366.

Dow, Sheila. 2009. History of Thought and Methodology in Pluralist Economics Education. *International Review of Economics Education* 8 (2): 41–57.

Drakopoulos, Stavros A. 2016. Economic Crisis, Economics Methodology and the Scientific Ideal of Physics. *Journal of Philosophical Economics* X (1): 28–57.

Elzinga, K.G. 1992. The Eleven Principles of Economics. *Southern Economic Journal* 58 (4): 861–879.

Fölster, S. 1995. The Perils of Peer Review in Economics and Other Sciences. *Journal of Evolutionary Economics* 5: 43–57.

Fourcade, Marion, Etienne Ollion, and Yann Algan. 2015. The Superiority of Economists. *Journal of Economic Perspectives* 29 (1): 89–114.

Frank, Robert H., Ben Bernanke, and Louis Johnston. 2013. *Principles of Macroeconomics.* 5th ed. New York: McGraw-Hill/Irwin.

Frey, Bruno S. 2003. Publishing as prostitution? Choosing between one's own ideas and academic success. *Public Choice* 116: 205–223.

Frey, Bruno S., and Katja Rost. 2010. Do Rankings Reflect Research Quality? *Journal of Applied Economics* 13 (1): 1–38.

Fullbrook, Edward. 2003. The crisis in economics. *The Post-autistic Economics Movement: The First 600 Days*. London: Routledge.

Gans, Joshua S., and George B. Shepherd. 1994. How Are the Mighty Fallen: Rejected Classic Articles by Leading Economists. *Journal of Economic Perspectives* 8 (1): 165–179.

Gibson, John, David L. Anderson, and John Tressler. 2014. Which Journal Rankings Best Explain Academic Salaries?: Evidence from the University of California. *Economic Inquiry* 52 (4): 1322–1340.

Gloetzl, Florentin and Aigner, Ernest. 2017. Six Dimensions of Concentration in Economics: Scientometric Evidence from a Large-Scale Data Set. *Ecological Economic Papers* 15. Vienna: WU Vienna University of Economics and Business.

Gwartney, J.D., Stroup, R., Sobel, R.S., Macpherson, D.A. 2015. *Economics: private and public choice*. 15th edition. Stamford: Cengage Learning.

Hahn, F. 1992a. Answer to Backhouse: Yes. *Royal Economic Society Newsletter* 78: 3–5.

———. 1992b. *Should We Ignore Methodology* 77: 5.

Heise, Arne. 2016. 'Why Has Economics Turned Out This Way?' A Socio-Economic Note on the Explanation of Monism in Economics. *Journal of Philosophical Economics* X (1): 81–101.

Hodgson, Geoffrey M., and Harry Rothman. 1999. The Editors and Authors of Economics Journals: A Case of Institutional Oligopoly? *Economic Journal* 109: 165–186.

King, John E. 2013. A Case for Pluralism in Economics. *The Economic and Labour Relations Review* 24 (1): 17–31.

Klamer, Arjo, and Hendrik P. van Dalen. 2002. Attention and the Art of Scientific Publishing. *Journal of Economic Methodology* 9 (3): 289–315.

Klein, Daniel B. 2005. The Ph.D. Circle in Academic Economics. *Econ Journal Watch* 2 (1): 133–148.

Kovanis, Michail, Raphaël Porcher, Philippe Ravaud, and Ludovic Trinquart. 2016. Complex Systems Approach to Scientific Publication and Peer-Review System: Development of an Agent-Based Model Calibrated with Empirical Journal Data. *Scientometrics* 106 (2): 695–715.

Krugman, Paul R., and Robin Wells. 2015. *Economics*. 4th ed. New York, NY: Worth Publ.

Kuhn, Thomas. 2012. *The Structure of Scientific Revolutions*. Chicago: Chicago University Press, 50th Anniversary Edition.

Laband, David N., and Robert D. Tollison. 2003. Dry Holes in Economic Research. *Kyklos* 56 (2): 161–173.

Larivière, Vincent, and Yves Gingras. 2009. The Impact Factor's Matthew Effect: A Natural Experiment in Bibliometrics. *Journal of the American Society for Information Science and Technology* 79 (3): 635–649.

Lawson, Tony. 1994. Why are so many economists so opposed to methodology? *Journal of Economic Methodology* 1 (1): 105–134.

Lee, Frederic S. 2006. The Ranking Game, Class, and Scholarship in American Mainstream Economics. *Australasian Journal of Economics Education* 3 (1): 1–39.

Lee, Frederic S., and Bruce C. Cronin. 2010. Research Quality Rankings of Heterodox Economic Journals in a Contested Discipline. *American Journal of Economics and Sociology* 69 (5): 1409–1451.

Mankiw, N. Gregory. 2015. *Principles of Economics*. 8th edition. Boston: Cengage Learning.

Mankiw, Nicholas Gregory, and Mark P. Taylor. 2014. *Economics*. 3rd ed. Andover: Cengage Learning.

Mayer, Thomas. 1993. *Truth Versus Precision in Economics*. Cheltenham: Edward Elgar Publishing.

McPherson, Michael A., Myungsup Kim, Megan Dorman, and Nishelli Perera. 2013. Research Output at US Economics Departments. *Applied Economics Letters* 20 (9): 889–892.

Mearman, Andrew, Danielle Guizzo, Sebastian Berger. 2018. Whither political economy? Evaluating the CORE project as a response to calls for change in economics teaching. *Review of Political Economy* 30 (2): 241–259.

Medoff, Marshall H. 2006. Evidence of a Harvard and Chicago Matthew Effect. *Journal of Economic Methodology* 13 (4): 485–506.

Merton, Robert K. 1968. The Matthew Effect in Science. *Science* 159: 56–63.

Mirowski, Philip. 1984. Physics and the 'Marginalist Revolution'. *Cambridge Journal of Economics* 8 (4): 361–379.

———. 1989. *More heat than light*. New York: Cambridge University Press.

———. 1991. The when, the how and the why of mathematical expression in the history of economics analysis. *Journal of Economic Perspectives* 5 (1): 145–157.

Neff, Bryan D., and Julian D. Olden. 2006. Is Peer Review a Game of Chance? *BioScience* 56 (5): 333–340.

O'Sullivan, Patrick. 2019. Economists' Personal Responsibility and Ethics. In *The Ethical Formation of Economists*, ed. Ioana Negru and Wilfred Dolfsma, 44–60. New York: Routledge.

Osterloh, Margit, and Bruno S. Frey. 2015. Ranking Games. *Evaluation Review* 39 (1): 102–129.

Oswald, Andrew J. 2007. An Examination of the Reliability of Prestigious Scholarly Journals: Evidence and Implications for Decision-Makers. *Economica* 74: 21–31.

Peters, Douglas P., and Stephen J. Ceci. 1982. Peer-review Practices of Psychological Journals: The Fate of Published Articles, Submitted Again. *Behavioral and Brain Sciences* 5: 187–255.

Reis, Ricardo. 2018. Is Something Really Wrong with Macroeconomics? *Oxford Review of Economic Policy* 34 (1–2): 132–155.

Ritzberger, Klaus. 2008. A Ranking of Journals in Economics and Related Fields. *German Economic Review* 8 (4): 402–430.

Seglen, Per O. 1992. The Skewness of Science. *Journal of the American Society for Information Science* 43 (9): 628–638.

Siegfried, John J., Robin L. Bartlett, W. Lee Hansen, Allen C. Kelley, Donald N. McCloskey, and Thomas H. Tietenberg. 1991. The Status and Prospects of the Economics Major. *The Journal of Economic Education* 22 (3): 197.

Skousen, Mark. 1997. The Perseverance of Paul Samuelson's Economics. *Journal of Economic Perspectives* 11 (2): 137–152.

Sobkowicz, Pawel. 2015. Innovation Suppression and Clique Evolution in Peer-Review-Based, Competitive Research Funding Systems: An Agent-Based Model. *Journal of Artificial Societies and Social Simulation* 18 (2): 13.

Squazzoni, Flaminio, and Claudio Gandelli. 2012. Saint Matthew Strikes Again: An Agent-Based Model of Peer Review and the Scientific Community Structure. *Journal of Informetrics* 6 (2): 265–275.

———. 2013. Opening the Black-Box of Peer Review: An Agent-Based Model of Scientist Behaviour. *Journal of Artificial Societies and Social Simulation* 16 (2): 3.

Stigler, George Joseph. 1988. *Memoirs of an Unregulated Economist*. Chicago: The University of Chicago Press.

Thurner, S., and R. Hanel. 2011. Peer-Review in a World with Rational Scientists: Toward Selection of the Average. *The European Physical Journal B* 84 (4): 707–711.

Tol, Richard S.J. 2009. The Matthew Effect Defined and Tested for the 100 Most Prolific Economists. *Journal of the American Society for Information Science and Technology* 60 (2): 420–426.

———. 2013. The Matthew Effect for Cohorts of Economists. *Journal of Informetrics* 7 (2): 522–527.

Vatn, Arild. 2009. Cooperative behavior and institutions. *Journal of Socio-Economics* 38: 188–196.

Vines, David, and Samuel Wills. 2018. The Rebuilding Macroeconomic Theory Project: An Analytical Assessment. *Oxford Review of Economic Policy* 34 (1–2): 1–42.

# CHAPTER 6

# Climate Change and Responsibility

**Abstract** Roos and Hoffart provide a justification for their call for more research more pluralism in climate economics. They argue that economists have both a social and a professional academic responsibility to address climate change. Starting with the responsibility of science in general, these responsibilities are derived from the goals of science. The chapter offers a much-needed discussion on the role and responsibility of economists, which the authors link to a general crisis of responsibility associated with climate change. Considering the urgency of climate change, it is problematic to deny these responsibilities. The general practice of mainstream economics results in serious conflicts with economists' academic responsibility and points to the bad state of neoclassical economics concerning the good practice of science.

**Keywords** Responsibility crisis • Practice of science • Openness • Economists' responsibilities • Social responsibility

## 6.1    Introduction

In this chapter, we justify our call for more economic research on climate change and more pluralism in climate economics that follows from our quantitative and qualitative critique (see Chaps. 2 and 4). We answer the

121
M. Roos, F. M. Hoffart, *Climate Economics*, Palgrave Studies in Sustainability, Environment and Macroeconomics, https://doi.org/10.1007/978-3-030-48423-1_6

questions *if* economists should address climate change in their research and *why* they should act accordingly and care about critique on climate economics. We do so by referring to the responsibilities of scientists, which we assume to also apply to economists.

Our main proposition is twofold. First, we argue that academic economists have a responsibility *qua* scientists to address climate change in their research in a way that contributes to the mitigation of and adaptation to climate change. Second, if economists do not act accordingly, it has implication for real life and the science of economics. To be more precise, denying a special responsibility means to not make full use of the powerful toolkit and expertise economists have to contribute to the mitigation of climate change. Considering the urgency of climate change, it is problematic to ignore the special responsibility. For the science of economics, ignoring the special responsibility reveals serious responsibility conflicts related to good practice of science.

By reflecting on the responsibilities of economists, we intend, on the one hand, to explicitly discuss the normative foundation, which is too often left out in economics (see Sect. 4.3.5 in Chap. 4). On the other hand, we aim to bring back questions concerning the philosophy of science and the self-understanding of economists to the discipline of economics (see Sect. 5.4 in Chap. 5). We welcome opponents' critique as a revival of a lively discussion on related issues in economics that were for too long neglected and labelled as undesirable. Although questions of ethics in economics are not new, we follow Stern and think that "discussions of responsibility and ethics are of great importance", especially when applied to economic climate policy (Stern 2011, pp. 8–9).

In Sect. 6.2, we link our critique on climate economics to a general crisis of responsibility that manifests itself in different degrees of responsibility denial related to climate change. We show that responsibility is not only relevant to justify our claim but also an issue in society and for the discipline of economics. In Sect. 6.3, we approach the question of economists' responsibility from a general perspective, defining the responsibility of science and specifying related responsibilities. We discuss why scientists have both a professional academic responsibility and a social responsibility. While the argumentation is rooted in the ethics in science and based on the goals of science, the understanding of responsibility evolved over time and relates to the development of science. Section 6.4 is devoted to the special responsibility of economics and is linked to economists' special role in society. First, we show that economists have a responsibility to address climate change, which can be derived from both types of scientists' responsibilities, namely the professional academic responsibility and social responsibility (Sect. 6.4.2). Second, we identify responsibility conflicts that result from the actual practice of mainstream climate economics that contradicts the good practice of science. Discussing conflicts with economists' academic responsibilities is not only relevant in the context of climate change but draws a general picture on the state of neoclassical economics.

When talking about economics, we refer to the science of economics taught and practised at academic institutions and to economists who work at those institutions. Defining economics in this way is based on the understanding that economics as a science is no entity by itself and cannot exist without the scientists. This aspect is especially relevant for assigning responsibility. We limit our attention to academic economists as opposed to practical economists employed in the public or private sector. The crucial point and a prerequisite for assigning responsibility is that academic economists, in contrast to practical economists, are free in their research and not obliged to vested interest. Freedom to choose is an important aspect of moral agency that is necessary for the fundamental responsibility of human beings, which serves as the basis for scientists' and thus economists' responsibility. We follow Douglas (2003, p. 59), who emphasises that scientists are "generally capable moral agents" who are not under threat or force. "Responsibility of scientists hinges on issues particular to professional boundaries and knowledge production". Interestingly, academic economists that prescribe themselves to positivism are also not excluded: "[D]espite a widespread tendency to denigrate and to expel moral considerations and ethical discourse from Economics as a subject over the past seventy years, economists inexorably remain free to choose and thus personally responsible for their choices and for their actions" (O'Sullivan 2019, p. 45).

## 6.2   Climate Change as a Responsibility Crisis

The concept of responsibility is not only relevant to justify our plea for more economic research on climate change and pluralism in economics. It is also discussed both in society and academia and is related to the lack of action associated with climate change mitigation. Especially the denial of responsibility is one problem in this context. There is a crisis of responsibility, which does not stop at the borders of society, but also affects the science of economics. To better understand the missing effort to fight climate change, we approach the question of responsibility in a broader way and link it to the science of economics.

When reading the recent news, it becomes clear that climate change is more than a change in the earth's climate, despite the term's literal meaning. Newspapers talk about the *climate crisis* and, more recently, scientists jointly warned that the "Earth is facing a climate emergency" (Ripple et al. 2020, p. 8). Climate change goes beyond a crisis of climate (e.g. Neubauer and Repenning 2019), in the sense of an environmental and ecological crisis. It is also a ....

- *Crisis of the economy*: How to transform the economy for deep decarbonisation?
- *Crisis of communication*: How to reach people who doubt climate science?
- *Crisis of justice*: How to mitigate climate change in a just way?
- *Crisis of energy*: How to change to low-carbon energy and address energy poverty?
- *Crisis of morality*: What can be new narratives for a good life in the twenty-first century?
- *Crisis of global politics*: How to unite all countries to implement the Paris Agreement?
- *Crisis of responsibility*: Who is responsible to mitigate climate change and to what degree?

This list can be continued and illustrates how far-reaching and multidimensional the consequences of climate change are. As portrayed in the Introduction, we think the lack of action to fight climate change is not so much a scientific but mostly a societal problem, since the solution is trivial and clear: we need to stop emitting greenhouse gases (GHG) into the atmosphere. However, how to stop emissions without crashing our societal systems such as the financial or the energy systems and how to allocate responsibilities in the required process of transformation is unclear. The lack of action seems to be related to a crisis of responsibility, which has different dimensions. First, it is not straightforward to assign responsibilities in the case of climate change, which lies in the nature of ethical considerations and represents the theoretical dimension. Second, even if it was theoretically possible to assign the responsibility to the *right* agents in a *just* way, this does, from a practical perspective, not imply that those agents accept their responsibility and act accordingly. This argument holds analogously to the search for the right price of carbon, often understood as a panacea to mitigate climate change (see Chap. 3).

### 6.2.1  Responsibility in the Context of Climate Change

Responsibility in climate change is very complex. The following discussion of (mostly) philosophical reasoning aims at illuminating this complexity. Article 3 of the Paris Agreement on Climate Change talks about the "principle of common but differentiated responsibilities" (UNFCCC 2015, Article 3). It indicates that there exist different principles, subjects and understandings of responsibility. Some argue that the ambiguity of responsibilities hinders adaptation measures (e.g. Juhola 2019). It is not surprising that people stay away from reflecting and taking responsibility, if responsibility and climate change is a complex and controversial topic. This reasoning might also explain why some economists tend to refrain from the responsibility to address climate change in their research.

In academia, debates on responsibility and climate change take place in at least three different fields, namely climate ethics, environmental management and environmental economics, which rarely interact (Doorn 2017). Hence, it is worthwhile to look at what responsibility means in the context of climate change and why it is difficult to assign responsibility, which bring us to the first dimension of the responsibility crisis.

Responsibility is relevant when talking about climate change mitigation and adaptation. Two questions which are interlinked are important: Who is responsible for causing anthropogenic climate change? Who is responsible for combatting climate change? While the first question refers to historical GHG emissions from the retrospective, the second is a prospective view into the future. Although these questions are addressed by different disciplines, the relation between responsibility and climate change is mainly a philosophical question and challenges the standard way of moral thinking (Gardiner 2011).

The challenge becomes clear when trying to answer the first question of *who is responsible for causing climate change*. In the traditional way, responsibility is assigned from a backward-looking perspective. It is related to blameworthiness and based on the legal concept of responsibility. In the traditional liability model, for example, to be morally responsible means that an agent is blameworthy for his/her action and the harmful consequences he/she directly caused (Young 2011). Since this way of assigning responsibility applies to individuals, most research on climate responsibility is on individual agents' moral culpability for climate change (Scavenius 2018). However, focusing on individuals is problematic in the case of climate change. While the individuals' emissions contribute to the accumulation of GHGs in the atmosphere, it is the accumulated effect of many GHG emissions that causes global warming and related harm (Steigleder 2018). It implies that it is problematic to assign responsibility in the traditional way to individuals, which is explained, for example, by Sinnott-Armstrong (2010). He demonstrates with the example of joyriding a fuel-guzzling car that it is not morally wrong to do so. Assigning responsibility for climate change to individuals based on their emissions is not possible, since the criteria of the traditional liability model are not met. GHG emissions of an individual do not contribute in a way to global warming that makes a significant difference. While emissions for an individual are not sufficient to cause global warming, not driving does not stop global warming either. It does neither imply that individuals are relieved from any responsibility nor that there are no other reasons for individuals to care about their emissions. Rather, the example shows difficulties that arise when assigning responsibility in the traditional way.

This aspect is important, since the responsibility for historical emission, and thus an agent's contribution to climate change, is often taken as the starting point to answer the second question of *who is responsible to fight climate change*. Alongside the so-called *polluter-pays-principles*, which catches up on the idea of past emissions, the *beneficiary-pays-principle* and the *ability-pays-principle* are discussed in academia and politics (e.g. Caney 2005). The latter two principles show that it is also possible to justify climate change mitigation without referring to historical emissions. While, according to the beneficiary-pays-principle, those who benefited the most from historical emissions should take most of the burden, the burden should be distributed based on people's capacity in the ability-pays-principle. Our main point here is that taking into account that in theory it is very difficult to assign responsibility in the case of climate change, the lack of action in reality is not surprising. However, it would be wrong to assume an intention to justify the lack of action.

### 6.2.2     Climate Change and Responsibility Denial

Lichtenberg (2010) re-defines the concept of harm related to global phenomena as a so-called *new harm*. According to her, it is not only necessary to rethink the concept of responsibility, but also the concept of harm. She points out that most people do not feel guilty when contributing to a new harm, such as climate change. Since an action, such as emitting GHGs, that contributes to a new harm does not cause an immediate and visible effect, in contrast to for example murder, people tend to ignore feelings of guilt or shame.

This might explain why people tend to ignore responsibility, which seems to apply to some degree to many economists. It leads to the second dimension of the crisis of responsibility, which relates to the denial of climate change. Compared to the first dimension, denial of climate change is no theoretical problem but a practical one. It comes in different degrees or, according to Cohen (2013), at three levels and has implications for the understanding of responsibility. Although the number of people who deny the existence of climate change at all (level 1: literal denial) or its anthropogenic causation (level 2: interpretative denial) decreases, there are still those who do not deny climate change itself but the responsibility to fight climate change (level 3: implicatory denial). This attitude is directed towards the implications resulting from the acceptance of climate change. Level 3 climate change deniers accept the reality of climate change but deny that we can and should do something. As a form of non-participation, they fail to translate their beliefs into action (Cohen 2013; Norgaard 2006).

In line with the idea of different levels of climate denial, we distinguish between different forms of *responsibility denial* based on the third level of climate denial. Climate change deniers of the third level do not deny the reality of climate change, but negate related responsibility. The first level of responsibility denial (level 3.1) refers to those people who deny that there exists any responsibility to mitigate climate change at all. At the second level, people only deny their own responsibility (level 3.2). The latter group of responsibility deniers do not deny responsibility in general but shift the responsibility to other individuals or actors, such as politicians, companies or future generations. According to them, it is others who are in charge to combat climate change due to different reasons. The third group of responsibility deniers (level 3.3) comprises those people who, in contrast to the former two types, do not deny their own responsibility but are not willing to act accordingly. Reif and Dahm (2017) describe the problem as a lack of willingness to take over responsibility. Sombetzki (2014) identifies the phenomenon of a so-called *Verantwortungsflucht*[1]

---

[1] English: escape from responsibility.

she observes in today's globalised world. It manifests itself in a tendency of people to not only ignoring their responsibility, but actively trying to retreat from it.

Jamieson (1992, p. 149) boils down the problem of the responsibility crisis: "Today we face the possibility that the global environment may be destroyed, yet no one will [or wants to] be responsible. This is a new problem". Similarly, Gardiner (2011) asks in his article of the same title *Is no one responsible for global environmental tragedy?* The crisis of responsibility is caused by a denial of responsibility. In Sect. 6.4, we will discuss about the special responsibility of economics in the context of climate change and provide concrete examples why economics are also affected.

### 6.2.3    Problematisation of Economists' Responsibility

We have demonstrated that the responsibility crisis has a theoretical and practical dimension and applies to different parts of society. Due to our focus on academic economists, we concentrate on the responsibility of science. Reflecting about responsibility in the context of economics requires elaborating on the role of science in society. We are aware that the discussion about the role of science is not a new but rather an old topic (e.g. Russell 1960; Jonas 1985; Douglas 2003).

However, pleas from new scientists' initiatives show that questions about the role and responsibility of scientists are more relevant than ever. The initiative Scientists for future (S4F), for example, talks about *responsible research* and assigns to itself a social responsibility: "As people who are familiar with scientific work and deeply concerned about the current developments, we consider it as our social responsibility to point out the consequences of inadequate action" (Hagedorn et al. 2019, p. 81). For them, "[r]emaining neutral and silent about our established state of knowledge on global environmental change would be a violation of our professional responsibilities towards our societies." (Hagedorn et al. 2019, p. 85). Similarly, the scientists who warned about a climate emergency argue having a "moral obligation to clearly warn humanity of any catastrophic threat and to 'tell it like it is'" (Ripple et al. 2020). These scientists are aware of possible critiques that might arise from their public positioning and accept that "[c]onflicts between scientific neutrality and political influence cannot always be avoided" (Scientists for Future 2019). These pleas highlight that the topic of scientists' responsibility is currently more relevant than ever before in a twofold sense. First, the fact that these scientists assign to themselves a special responsibility related to climate change shows that the topic is of special concern for them. Second, they reveal a need to specify and justify the related responsibility that we aim to satisfy.

We think that the question of responsibility is especially relevant for the science of economics and follow Dolfsma and Negru (2019, p. i) who argue that "[e]conomists' role in society has always been an uneasy." Different voices provide first indicators, that responsibility in economics is problematic in different ways.

Economists are accused of not fulfilling their responsibility or behaving irresponsibly. DeMartino (2014, p. 88) thinks that "if you are not an economist, you would be forgiven for thinking that the awesome responsibilities that come with this influence keep economists up at night". Morgan (2019, p. 147) formulates it more provocatively, saying that "[r]esponsibility is a problematic concept for economists. Insofar as economists have become society's go-to voice [...] they have arguably *accepted* responsibility, and yet economists do not clearly *take* responsibility."

Especially in the context of the global financial crisis, the responsibility of economics is problematised and at the heart of discussions. Economists are criticised for bearing at least some responsibility for the financial crisis (e.g. Schneider and Kirchgässner 2009). More extreme voices think that the crisis is caused by a systematic failure of academic economics (e.g. Colander et al. (2009). Heise (2009) describes economics even as a *toxic science* in the context of the financial crisis.

The way economists reflect about responsibility can also be framed as problematic. Morgan (2019) points out that discussions about economists' responsibility for the financial crisis concentrate on irresponsibility by pointing to questionable advice or inadequate research approaches. To us, reflecting about irresponsibility has two implications. First, it underlines that questions of responsibility inevitably belong to ethics in economics. Thinking about responsibility with the intention to avoid irresponsibility instead of fostering responsibility reveals something about the priority given to ethics in economics. It is not without reason that DeMartino (2011) calls for an economists' oath and hopes to "generate the new field of professional economic ethics" (DeMartino and McCloskey 2016, p. 4). By doing so, he started a controversial discussion on the need for a code of conduct for economists (e.g. Dow 2013; DeMartino 2013, 2014; DeMartino and McCloskey 2016; Krueger 2017; DeMartino and McCloskey 2018). Second, it hints to what Morgan (2019) explicitly laments—economists are neither trained nor supported to reflect about their responsibility. Not reflecting on responsibility might also explain why the science of economics might be affected by the responsibility crisis.

Many mainstream economists reject to be held responsible for the implication of their research and policy advice. A common argument is that it is not them but others, namely politicians, who translate economic theories and advice into policies (Roos 2016). Economists sometimes recommend policy measures that would increase overall efficiency but have negative distributional consequences for some groups. An example is the standard plea for free trade. Typically, economists acknowledge that trade liberalisation generates winners and losers, but they emphasise that the overall size of the pie will grow such that the winners can compensate the losers. How this compensation might work in the political practice is neither part of the economics analysis nor of their policy recommendations. This example is linked to DeMartino (2016) who introduces the term *econogenic harm*. This harm results from the practice of economics when trying to do good but neglecting the normative consequences.

O'Sullivan (2019) links *econogenic harm* with scientists' responsibility. For him, the neglect of ethical considerations related to *econogenic harm* has negative implications for the economists' responsibility. The mathematisation of the discipline in the twentieth century explains the withdrawal from personal responsibility *qua* scientist. Especially Friedman and Friedman's (1953) problematic methodological claim of staying away from value judgements influenced many economics. It contributed to a "serious erosion of any sense of personal ethical responsibility among professional economists. There has been an effective abdication of moral responsibility for the consequences of their policy advice by large swathes of mainstream neoclassical economists" (O'Sullivan 2019, p. 47). O'Sullivan (2019) certifies economists a certain insensitivity for moral considerations both in theorising and policy advice, which leads to an insensitivity in terms of responsibility.

Although we point to problems of responsibility in economics, we do not think that most economists are, in general, irresponsible or lack any sense of responsibility. It is, though, hard to deny that economists feel uncomfortable reflecting responsibility and are affected by a crisis of responsibility.

## 6.3   General Responsibilities of Scientists

To discuss economists' responsibilities in the light of the responsibility crisis, we specify on the general question of responsibility in science. This general responsibility is the normative foundation of our call for more economic research on climate change and pluralism in climate economics.

### 6.3.1   Foundation of Scientists' Responsibilities

It is hardly deniable that science has an influence on society. Verhoog (1981, p. 582) talks about scientists' different responsibilities in a "society which is so much affected by science". Douglas (2003, p. 66) refers to the "awesome power of science to change both our world, our lives". He asks what society should expect of scientists qua scientists. This question is becoming more urgent due to the increasing importance of science he observes over the last decades.

More than a decade after his writing, this question did not lose its importance facing the urgency of climate change; the contrary is the case. On the one hand, society puts great hope on science when it comes to climate change mitigation, especially the development of technical solutions to stop emissions. On the other hand, scientists' movements, such as S4F, are not only welcomed but also criticised for their interaction with politics. Reacting to questions posed by Douglas (2003), many scientific institutions developed ethical standards, such as codes of ethics, including the aspect of responsibility. Document *Science Agenda—Framework for Action*, for example, recommends, in 1999, that "[t]he ethics and responsibility of science should be an integral part of the education and training of all scientists" (Cetto 2000, p. 482). It implies that ethics in science and responsibility cannot be separated.

According to Hansen (2006), it is not clear what responsibility associated with science means. The literature provides no clear definition. The term *responsibility* has several meanings and allows for different interpretations. In a general sense, responsibility has two dimensions. A person can be responsible for something (*for what?*) and responsible towards somebody (*to whom?*). The former can be interpreted both in the sense of *being the cause* of some consequence and being responsible in the sense of *an obligation* related to a specific role in society (Verhoog 1981).

Some argue for a link between ethical standards, responsibility of science and the goals of science (e.g. Resnik 2005; Longino 1990). According to Verhoog (1981), responsibilities of scientists can be formulated in the form of norms or rules by groups or institutions and are based on ethical standards. Therefore, scientists' responsibilities belong to ethical standards in science. For Evers (2001), science's ethical standards are based on ethical values of the society that are relevant for the everyday life, such as fairness or respect. Resnik (2005) argues that ethical standards in science are grounded on the profession's goals and have two conceptional foundations: science and morality. It implies that ethical standards in science, and thus responsibilities, should promote scientific goals (science) and should not violate moral standards (morality). In the same vein, Douglas (2003) elaborates that the scientists' role responsibility should be built around the goal of science. These reflections reveal, first, that the goal of science is a good starting point to approach scientists' responsibility. Second, they indicate the existence of different types of responsibility, as the term *role* responsibility implies. Third, they point to the unclear addressee of science's responsibility, which can be both the individual scientist and the collective of scientists.

Referring to the goal of science, Resnik (2005) differentiates between the goal of traditional, academic science and the goal of science practiced in a non-academic context. The difference is important for the freedom of research and constraints on the latter. Resnik (1996) elaborates that there is no single goal of traditional, academic science, but two categories of goals, namely epistemic and practical goals. The former includes research that expands human knowledge—which is often described as the search for truth.[2] The latter is concerned with the application of knowledge to find solutions to practical problems—with the aim to improve human life (Resnik 2005).

For Kitcher (1993), it is crucial to distinguish also between science's goals and scientists' goals. Science's goals can be described as the goals of the profession of science. In contrast, scientists' goals represent individual goals of scientists. While they can overlap, scientists' goals, for example, employment or prestige, usually go beyond those of science. Similarly, intrinsic goals of scientists, such as research for truth (Agazzi 2014) can be differentiated from extrinsic counterparts, such as career or money (Alai 2016). Scientists' extrinsic goals are independent of the scientific professions and apply to all kinds of professions. Hence, they should neither influence nor justify ethical scientific standards.

---

[2] Or reliable knowledge.

Is it the individual scientist or the collective of scientists such as scientific institutions? We follow Hansen (2006), who argues for the responsibility of both. He developed an analytical model of three interacting levels, that shows the interactions between (I) ethical values of science, (II) science's social mechanism and institutions and (III) scientists' individual responsibility. On the first normative level, ethical principles and responsibilities are defined by the collective. These principles influence the formation of scientific institutions (level II) and the behaviour of scientists (level III). The dividing line between the individual scientist and the collective of scientists is blurry. For the sake of the argument, we emphasise the responsibility of individual scientists, which is considered to be of special importance according to the *Standing Committee on Responsibility and Ethics*: "[t]he ethical responsibility of the scientific community is ultimately borne by the individual scientists. [...] the ethical awareness of the individual scientist is of utmost importance" (ICSU 1999). Again, the literature on responsibility in the context of science is not clear. Similar but different terms and wordings are used, which requires clarification.

### 6.3.2    Overview of Scientists' Responsibilities

Table 6.1 is based on the previous reflections and represents an attempt to define and classify scientists' responsibilities. Following Shrader-Frechette (1994), we define scientists' responsibility as the responsibility scientists have for the research. To specify, it refers both to the conduct of research (A) and to the implications of research (B). The former is related to the process of science and will thus be called *professional academic responsibility*. The latter relates to the product of science and will be called *social responsibility*. For Bird (2014, p. 169), "social responsibility is the other side of the coin of the responsible conduct of research". Sakharov (1981, p. 184) agrees and points out that "[his] view of the situation of scientists in the contemporary world has convinced [him] that they have special professional and social responsibilities."

The two types of responsibility can be described as role responsibilities. Role responsibilities are rooted in a person's position in society, which explains the naming *professional academic responsibility* (Douglas 2003). The Committee on Science et al. (2009) propose the following three sets of obligations: (1) responsibility for oneself as scientist, (2) responsibility for society, (3) responsibility for the scientific community. We apply these to the responsibility in science and complete the set with a fourth

**Table 6.1**   Overview of scientists' responsibilities

| Responsibility of scientists | | | |
|---|---|---|---|
| **(A) professional academic responsibilities** | | **(B) social responsibilities** | |
| Function:<br>to serve the epistemic goal<br>of science | Foundation:<br>role responsibility of<br>scientists | Function:<br>to serve the practical goal<br>of science | Foundation:<br>based on moral<br>responsibility of citizens |
| **(A1) Responsibility to conduct reliable research**<br><br>1.1 Honesty (no misrepresentation of data)<br>1.2 Carefulness (minimize errors)<br>1.3 Avoidance of vested interests<br>1.4 Sharing of data, method, tools results<br>1.5 Obey law & moral standards (e.g. in laboratory)<br>1.6 Fair behaviour & respect<br>1.7 Efficient use of resources and research money | | **(B1) Responsibility to care for the consequences of research**<br><br>1.1 Counteraction of possible misuse &<br>    misinterpretation of research<br>1.2 Reflection about limitations and foreseeable<br>    impacts of their work<br>1.3 Reflection & discussions about the adequate<br>    application for addressing societal issues | |
| ≙ Responsibility for<br>oneself as a scientist | ≙ Responsibility for the<br>scientific community | ≙ Responsibility for society | ≙ Responsibility for oneself as<br>a citizen |
| **(A2) Responsibility to teach and educate future scientists**<br><br>2.1 Teaching the whole discipline (variety & depth)<br>2.2 Reflections about the philosophy of science, and<br>    methodology | | **(B2) Responsibility to seek for relevance in science and to conduct socially valuable research**<br><br>2.1 Research on urgent & relevant topics (quantity)<br>2.2 Focus on usefulness & applicability | |
| ≙ Responsibility for the scientific community | | ≙ Responsibility for society | ≙ Responsibility for oneself as<br>a citizen |
| **(A3) Responsibility to participate in academic discussions**<br><br>3.1 about own research, which includes<br>    3.1.1 openness (intra- & interdisciplinary)<br>    3.1.2 a response to criticism<br>    3.1.3 reflections about the methodology,<br>        normativity, philosophy of science<br><br>3.2 about research of others, which includes<br>    3.2.1 to provide feedback, unbiased review<br>    3.2.2 a general awareness & interest | | **(B3) Responsibility to communicate and advice the public and governments**<br><br>3.1 Communication with public & governments<br>3.2 Provision of advice for public policy | |
| ≙ Responsibility for<br>oneself as a scientist | ≙ Responsibility for the<br>scientific community | ≙ Responsibility for society | ≙ Responsibility for oneself as<br>a citizen |

Source: Authors' own contribution

obligation: (4) responsibility for oneself as a citizen. These specific responsibilities are based on a broad range of literature and are not exhaustive.

The professional academic responsibility to conduct reliable research (A1) is widely accepted. Related violations are typically punished directly or indirectly. For example, misrepresentation of data (1.1), violations of the law (1.5) or wasteful use of research money can result in legal action or sanctions by universities or funding institutions. Lack of care (1.2), refusal to share data or tools (1.4) or disrespectful behaviour in seminars or at conferences (1.6) might result in social sanctions and loss of reputation among the peers.

The responsibility to teach and educate future scientists (A2) is also uncontroversial, at least in the sense that it belongs to the job of academics to teach. However, the responsibility does not only refer to the *fact that* teaching is necessary but also to the *way how* this is done. In order to reproduce well-functioning future scientific community, variety in research (2.1) and self-reflection on the role of scientists and scientific methodology (2.2) should be transmitted to future scientists. As we will discuss in Sect. 6.4, economics does not always take this responsibility.

Similarly, the responsibility to participate in academic discussions (A3) is accepted by all researchers, because all scientists permanently read and respond to the work of others. As in the case of teaching, a problematic issue is not *that* this responsibility exists, but *how* it is realised in everyday scientific practice. We will also discuss this point for economics in Sect. 6.4.

While the professional academic responsibilities in general are widely accepted, social responsibilities are not. Therefore, and because they are of special importance in the context of climate change, we discuss social responsibilities in more detail in the following subsection.

### 6.3.3   Discussion of Social Responsibilities

The "relationship between science and society has changed fundamentally in modern knowledge societies over the past decades" (Schneidewind et al. 2016, p. 4). The related discussion dates back to the beginning of science. It is characterised by the search for the adequate balance between professional autonomy, that is, freedom of science, and general moral responsibility of humans (Douglas 2003). Ziman (1998) is wondering "[w]hy are scientists now expected to be so much more ethically sensitive than they used to be?"

Similarly, to Verhoog (1981), understanding scientists' responsibilities is related to the institutional development of science. He defines three stages of institutionalisation of science, namely *amateur science, academic science and industrial science*, which have implications for the responsibility in science. Concentrating on the second and third phases, he elaborates that academic science started in the nineteenth century and originated in the Germany University reform. The academic ethos of this time can be described as an Aristotelian ideal of highlighting the intrinsic value of scientific knowledge. Scientific knowledge is gained *for its own sake* and does not aim for relevance and applicability, but for *Bildung* (engl: education). With the specialisation of research in different disciplines that took place in the second half of the nineteenth century, the ethos changed. The idea to expand scientific knowledge gradually replaced the traditional ethos of knowledge for the sake of education as the dominating goal of science. In the phase of industrial science, starting in the twentieth century, a shift towards a more practical, utility-oriented science took place in many disciplines. Utility became an additional goal of science. With increasing influence on society through technological developments, known as *scientification*, some called for an extension of responsibility (Verhoog 1981).

Verhoog (1981) identified three major positions in the discussion about the responsibility, namely *neutrality in science*, the *scientistic view* and the *critical-interactionist view*. While the first two were rather traditional, widely shared views, the latter was a minority position. From a neutrality of science view, the only goal of science and thus responsibility of scientists is to generate scientific knowledge for its own sake. In addition, scientists are responsible to follow scientific norms, but also have responsibilities related to the broader societal context in which their research is embedded. Society itself is responsible for the use and implication of scientific knowledge. For proponents of the *scientistic view*, knowledge is not evaluatively neutral when it comes to daily life. They argue that science is applicable to daily life and that science should focus more on societal aspects. It is the responsibility of scientists to promote the rational way of thinking. A technocrat society is desirable, since science can build the rational base for society, which is known as positivistic conception of science. Verhoog proposes the critical-interactionist view, which rejects the positivists' view of scientific thinking as the adequate way to understand society. Consequently, there is no difference between the responsibility of scientist *qua* scientists and *qua* citizens. The value of knowledge acquisition (Aristotelian ideal) cannot be separated from its value for society (Baconian ideal). Scientific knowledge "requires critical reflection upon the social system in which science is applied and upon the purpose for which science is used. [...] For this reason, it belongs to the social responsibility of the scientist and scientific technologist to take part in this discussion" (Verhoog 1981, p. 589).

Although Verhoog's reflections might not display the recent development, his work captures still existing views. It describes what Beckwith and Huang (2005, p. 148) explicitly state, namely that the "the sense of social responsibility in science has emerged from time to time in spite of the fact that scientists were not prepared by their training to think about these issues. Their activism was stimulated by crises". Referring to the implication of crisis raises the thought that the climate crisis might also stimulate a new sense of social responsibility among scientists and economists.

In recent years, the literature on scientists' social responsibility is growing and is enriched by work from policymakers (e.g. Stilgoe et al. 2013). As Glerup and Horst (2014) explain, the debate is shifting to the aspect of *new governance of science and* relates to discussions about the responsible way of practicing and governing science. To capture this new development, Glerup and Horst (2014) analysed 263 journal articles on responsibility in science. They identified four related understandings of social responsibility of science, which form a 2 × 2 matrix. While the first axis displays the regulation of science (internal or external), the second axis links responsibility to the process or the result of science:

> *The Reflexivity and Demarcation rationalities both advocate internal regulation of science but, while the Reflexivity rationality insists that the responsibility of science is to strive for outcomes that can work as solutions to society's problems, the Demarcation rationality aims for a total separation between science and society in order to prevent social norms and values from biasing the otherwise objective production of knowledge.* (Glerup and Horst 2014, p. 35)

Comparing the four understandings of social responsibility highlights the unavoidable relation of scientists and society nowadays. As some still deny a social responsibility, which is crucial to argue for a responsibility to address climate change, it is worth discussing proponents' and opponents' arguments. In the literature, two main reasons are highlighted to support social responsibility in science: scientists' moral duties *qua* humans and obligation to do good if it is at reasonable costs (e.g. Douglas 2003); scientists' membership and role in society (e.g. Shrader-Frechette 1994). Based on these reasons, we refer to three types of argument that we summarise as the *argument of interdependence, society's need for science* and *power entails responsibility.*

The *argument of interdependence* highlights the special mutual relationship between science and society. Scientists have a social responsibility through being a member of society (Shrader-Frechette 1994). The relationship between science and society can be interpreted at least in four ways. First, science is embedded within society implying that it takes place in a social context. Accordingly, science can be understood as a social institution. Second, science interacts with society in the sense that it has societal consequences. Third, science depends financially on society, since it is largely subsidised and funded through public funds and taxes. Understood as undeserved gift, scientists have in return a responsibility towards the donor, namely society (Camenisch 1983). Finally, society leans on scientific research and advice. This is true, for example, for the provision of higher education.

The argument on *society's need for science* outlines in various areas as the grounds for social responsibility. The availability of knowledge is important for society and supports citizens in realising their right to free informed consent. Sharing scientific knowledge with society enables a fully informed decision and reduces the room for political arbitrariness (AAAS 1980). Additionally, scientists' advice and knowledge are needed to help to solve societal problems. Climate change is the most urgent societal problem falling into this category, where science can contribute to find mitigation measures and discuss its implications.

*Power entails responsibility* is the claim of the third argument. It is based on the idea that power, ability and knowledge includes responsibility (Shrader-Frechette 1994). Science has a direct and indirect influence on society through their political and intellectual power, which implies a certain responsibility towards society. The power is due to science's special knowledge and the monopoly position of scientists (Camenisch 1983). A greater knowledge compared to non-scientists also implies a greater responsibility (Edsall 1975). Analogously, "[t]he power of science simply makes the fulfilment of the [social] responsibility more urgent" (Douglas 2003, p. 66).

Some do not deny social responsibility, but think that academic responsibilities simply outweigh social responsibility. This perspective is in line with the scientistic view of responsibility and is related to the freedom of science movement in the 1940s. Bridgman (1947, p. 72), for example, describes social responsibility as a handicap and restriction of scientific freedom: "The challenge to the understanding of nature is a challenge to the utmost capacity in us. In accepting the challenge, man can dare to accept no handicaps. That is the reason that scientific freedom is essential, and that the artificial limitations of tools or subject matter are unthinkable". For Lübbe (1986, p. 82), scientists do not have the same responsibilities non-scientists have. They are released from some responsibilities and have "a morally unencumbered freedom from permanent pressure to moral self-reflection".

Douglas (2003) discusses why role responsibility (professional academic responsibility) does not outweigh scientists' general moral responsibility *qua* citizen and moral agent (social responsibility). He provides two main reasons which we agree on: First, the epistemic goal of science cannot outstrip all other values. Seeking for knowledge is no ultimate value and therefore not ranked above all other values. There are situations where knowledge acquisition has its limits. Parents, for example, would hardly agree to sacrifice their children and allow medical life-threatening experiments for the sake of science. Second, the influence and implication of knowledge for society, is hard to control. Scientists are best qualified to control their research. Delegating the responsibility to others would imply that "[s]cientists are left with a choice: either accept the burden of general responsibilities themselves, or lose much of their [...] authority in allowing others to take on the burden for them" (Douglas 2003, p. 60).

The discussion of the role of science in society continues in the recent times as well. In Germany, a lively debate between Uwe Schneidewind, current president of the Wuppertal Institute for Climate, Environment and Energy, and the former president of the Deutsche Forschungsgesellschaft[3] Peter Strohschneider took place. Schneidewind and Strohschneider discussed the concept of *transformative science* which calls for a new role of science in the modern knowledge societies and fosters related social responsibility. According to Schneidewind, research needs to reorientate towards the needs of society, because great societal challenges such as climate change require the support and advice of science (e.g. Schneidewind et al. 2016). His opponent is highly sceptical about this view and fears that transformative science implies a politicisation of science and a depolarisation of politics, that he considers undesirable (Strohschneider 2014).

---

[3] The main German research funding organisation.

## 6.4    The Special Responsibility of Economists

Our main proposition is twofold. First, we argue that economists have a responsibility to address climate change that is not automatically fulfilled through the mere existence of research on climate change. For this purpose, scientists' social responsibility is of special importance. Second, we argue that the current mainstream climate economics conflicts with scientists' general responsibility. We see several conflicts with the professional academic responsibilities concerning the good practice of science.

There are different ways to show why climate change is a subject matter for economics. One could refer to the mutual interaction and effects between the economy and the environment or the understanding of climate change as an externality. It is also possible to point to actual research to prove that environmental aspects are research matters of economics. Additionally, the existence of different schools of thoughts devoting their research to environmental aspects, such as climate economics or ecological economics (Chap. 4), supports our reasoning. The latest, since Nordhaus was honoured with the Nobel Prize for his work on climate change, environmental aspects are recognised as a subject matter of economics.

Opponents might defend current climate economics by saying that economists do their job and address climate change by referring to the research efforts to identify the social costs of carbon. They could explain that they fulfilled their duty to address climate change and found the answer to how to fight climate change, namely through market-based instrument, such as a carbon tax. According to this view, it is now up to the others such as politicians to translate their advice into practice.

The "Economists' Statement on Carbon Dividends" that was published in January 2019 in *The Wall Street Journal* is an example for this kind of reasoning. In the statement, 48 economists agreed that "[g]lobal climate change is a serious problem calling for immediate national action" (Akerlof et al. 2019). Furthermore, they recommend what should now be done to successfully translate their advice into practice. Formulations of this kind imply that the signatories think that their job is done by now and that others, namely politicians, need to act now. The initial 48 initiators are famous and renowned economists, such as the Nobel Laureates Amartya Sen and Daniel Kahneman or the former chair of the US Federal Reserve Janet Yellen. According to their homepage, 3559 economists followed and signed the statement (Climate Leadership Council 2019).

In March 1997, a group of economists published a similar statement, the "Economists' Statement on Climate Change", and proposed market-based policies as *the* measure to slow climate change. The statement which was, at this time, the most signed statement in the history of economics. It was published even before the Kyoto Climate Change Conference took place in the same year. The New York Times reported the economists' statement and highlighted that economists think that it is now up to the national governments to act (Passel 1997).

One could argue that economists should care about climate change because it is an economic subject matter. However, to care about the critique we raised in Chaps. 4 and 5 would not be required, since economists fulfilled their job and duty. In order to argue for more economic research on climate change and the use of the whole toolkit and expertise of economics associated with pluralism, requires a different kind of reasoning. For this purpose, we use scientists' responsibilities as a normative foundation for our claim and apply it to economists. It should be obvious that responsibilities which apply to all scientists hold for economists, too.

Economists have also a special role, at least among the social scientists, which implies a special responsibility. One way to justify this proposition is to use the responsibilities of science as a basis and to extrapolate to the responsibility of economists in the context of climate change. Discussing about normative questions allows for different approaches. It does also apply to our case of justifying a responsibility to address climate change. However, our focus is not so much on the approach itself. Instead, we intend to bring back discussions about normative questions to the science of economics which hardly take place but are considered undesirable (Chap. 5).

### 6.4.1  *The Special Role of Economists*

As the role of science in society is not only relevant for the questions of responsibility but "is a matter of the greatest importance […] and may well be determinative of the future course of civilization" (Bridgman 1947, p. 69), it is worth discussing the role of economists in society.

The role of economists is special in the sense that it is argued to be clear and undisputed for economists themselves. Although economists' role is arguably no easy one (Dolfsma and Negru 2019), for economists it seems to be clear, at least, when it comes to policy. According to Morgan (2019, p. 145), economists take it for granted that their role extends to the spheres of public policy and intervention. Contrary to other social sciences such as political science, "there is no discourse of economic deferment, no economics of silence. Intervention is, tacitly at least, considered intrinsic to economics, a right and a duty".

Their self-imposed, unquestioned right for recommendations on policy interventions is related to the dominating self-image of economists and how they are socialised. The majority of economists see themselves at the top of disciplines in the social sciences (Fourcade et al. 2015). Economists' perception is amplified by what they have become over time, namely important people to consult for advice. In this sense, economists' perception of their role in policy decision-making and resulting authority is confirmed by the public. It is not without reason that economics is described as the "world's most powerful profession" (DeMartino 2014, p. 89). It is claimed that the "importance of economics in our daily lives and well-being can hardly be overestimated" (Groot and van den Brink 2019, p. 141). To DeMartino (2014, p. 88), in "the social sciences, there is no other discipline whose practitioners enjoy the impact that economists routinely have today". Similarly, Hirschman and Berman (2014, p. 779) describe economic science as the "most politically influential social science".

While economists have a *feeling of superiority* and became "first port of call consultants [..] and the preferred public intellectual" (Morgan 2019, p. 146), some believe that their influence is small. Although they are consulted for advice, it is not them but others, namely policymakers, who decide and implement policies. However, there seems to be a certain desire amongst economists to be more influential and heard by the public. The signatories of the *Economists' Statement on Carbon Dividends* and the *Economists' Statement on Climate Change* can be named as examples. They believe in an improvement if their advice was to be considered.

Both economic theories and economists themselves are relevant for society and influential in direct and indirect ways.[4] Concerning the former, there are fields of economics that are more applicable and influential than basic economic research, which is not intended to be applied. But even basic research can, in a second step, be applied to the real economy. The idea of influential economics is in line with Callon (1998), who initiated the performativity thesis. It states that "economics does not describe an existing external 'economy', but [...] performs the economy, creating the phenomena it describes" (MacKenzie and Millo 2003, p. 108). Colander et al. (2009) support the performativity thesis by referring to the financial

---

[4] These arguments are developed based on earlier work from Roos (2016).

crisis and the co-responsibility of economists. Besides, the key role of the gross national product in policy and political discussion is a supporting example. Similarly, Faulhaber and Baumol (1988) describe economists as innovators. For them, economists overlook that their theories (in)directly cause numerous innovations in the industries, for example, peak-load pricing, while studying the economic impact of technological innovation.

Not only economic theories have a societal influence but also economists, because they give advice to decision-makers. A prominent example is Nicolas Stern, a renowned professor for economics at the London School of Economics and Political Science (LSE) and former World bank economist (LSE 2017). In 2005, he was asked by the UK Government for a report on climate change he published one year later (Stern 2007). What is notable about the 700-page report, which is surely neither the first nor last economic report on climate change, is that, of all thinkable scientists working on climate change, the government chose an economist to be in lead. This choice had implications for the focus of the report, since every scientific discipline introduces a particular point of view. "The report would have been differently framed if it had been headed by climate scientists or social policy experts. The economy would have been a subordinate issue rather than the priority through which all else was filtered" (Morgan 2019, p. 146). Economists shape the political discussion on how to mitigate climate change like only very few other disciplines do.

By providing policy advice, economists inevitably impose their worldview and values on society and politics. This is especially true for neoclassical economists. The fact that $CO_2$ pricing became *the* tool of choice to mitigate climate change can easily be attributed to a focus on market-based instruments and thus an influence of economics. By placing $CO_2$ pricing at the heart of the political discussion, "the world is governed by the philosophy, or ideology, of economic science as practiced by economists" (Groot and van den Brink 2019, p. 139).

Besides providing policy advice, economists are decision-makers themselves and play a major role in the policy process. Many central bank governors were university professors before they joined the central bank, for example, Ben Bernanke and Janet Yellen (U.S. Fed), Mervyn King (Bank of England), Wim Duisenberg and Mario Draghi (ECB). In many countries, economists hold the highest offices in governments, for example, George Shultz and Lawrence Summers as U.S. Secretary of the Treasury or Yanis Varoufakis as Greek Minister of Finance.

Also, academic economists working at universities are influential in a multiple sense. Economic professors educate their students and thus the future economists who will take over different positions in the public and private sectors. Mearman and McMaster Robert (2019) discuss about the influence of economists' training on their future jobs and their sense of ethics.

Economists get a lot of attention from media. Groot and van den Brink (2019) explain that economists are often approached by media to comment on a current economics topic and development, such as the banking crisis, the recession or the job market. Especially US economists appear frequently on media. By doing so, they not only partly form the public opinion but also "increase the relevance and status of the economic profession in the eyes of the general public" (Groot and van den Brink 2019, p. 137). In line with Hirschman and Berman (2014), we argue that such indirect ways of influence are at least as important as the direct way of influence, which should not be underestimated.

These reflections support the view that economists have a high societal influence. It also demonstrates that economists shape the political discussion on how to mitigate climate change like no other discipline in the social sciences. Based on the idea that more power implies more responsibility, we argue that economists do not have the same responsibility as scientists in general, but a special responsibility. It neither implies that economists have the most responsibility for addressing climate change, nor that other disciplines have none. It seems undisputable that the expertise of all scientists is needed to fight climate change.

### 6.4.2   Economists' Responsibility to Address Climate Change

The overall thesis of this book it that economists should do more research on climate change and do it in a different way associated with the current mainstream climate economics. Other economists, such as those active in the Econ4Future movement, agree. In their open letter, they formulate a similar statement: "We are living through a climate emergency and, among the many profound challenges that this presents, the situation demands that the discipline of economics takes a hard look at itself". According to the movement (2019, p. 1) "[e]mergencies do not call for incrementalism, they call for an intervention. If the discipline which dedicates itself to studying the economy cannot sufficiently engage in the economic transformation that the climate science requires, then who else can be expected to do this? The responsibility and the opportunity is ours." Econ4Future argue for a responsibility of economics in the case of climate change. Thus, providing justification and discussing the underlying normative foundation of this claim is highly relevant for the questions of responsibility.

The responsibility of academic economists to address climate change, which is also implicitly included in the statement of Econ4Future, can be derived from both professional academic responsibility and social responsibility of economists. One way to argue for this specific responsibility is to refer to different responsibilities of scientists and apply these to climate change. For this purpose, we refer to the following specific responsibilities, taken from Table 6.1.

- *Responsibility to participate in critical academic discourse (A3)*
- *Responsibility to care for the consequences of research (B1)*
- *Responsibility to seek for relevance in science and to conduct socially valuable research (B2)*

Referring to the social responsibilities of scientists and economists directly hints to a responsibility to address climate change. Reflecting on social responsibility (B1), it is hard to deny consequences of economic research, especially in the context of climate change. Economists generally assume that economic growth increases welfare because of expanding consumption possibilities. Economic growth, its determinants and the possibility of the government to promote it are central topics in (macro-) economics. Under the current conditions of production, economic growth is linked to rising GHG emissions. By promoting economic growth without considering planetary boundaries, neoclassical economics indirectly provokes an increase in emissions and thus has indirect negative environmental effects. Besides, economics and especially macroeconomics has a strong focus on economic policy and involves identifying optimal policies, such as the optimal amount of pollutions. Ackerman (2008, p. 326) highlights that "[e]conomists' [...] conclusions about climate change echo throughout the public debate; economic analysis has a major impact on the decisions that politicians and governments are willing to take". These examples underline that economists' research is not without consequences in the case of climate change. Thus, the responsibility to care for consequences of research does include considering environmental issues and allows justifying a responsibility to address climate change.

Social responsibility (B2) calls for relevance in research. That climate change is indeed relevant for society should be undisputable. Hence, economic research on climate change is, without doubt, socially valuable and urgently needed. As explained in Chap. 4, there are a lot of different aspects where economic research can make a valuable contribution in the context of climate change. Due to their high influence on society, which is related to a special need of the public for advice, Ackerman (2013) sees a potential of climate economics to bridge science (in theory) and policy (in practice). This is especially true for plural climate economics, which does not only focus on the identification of policy but also studies the implementation and other *untouched* aspects. According to Oswald and Stern (2019, p. 5), "economics can and should play a fundamental role in guiding the policy framework" that is necessary to mitigate climate change. Besides, climate change is not only relevant for society but also for the science of economics itself. Climate change issues are no additional topic for economists they would normally not think about but are part of economic research. Thus, the responsibility to seek for relevance in science does include addressing climate change and allows calling for a responsibility to do so.

These examples highlight in different ways how to derive the responsibility to address climate change from economists' social responsibilities. But even if critics reject a social responsibility in general, responsibilities for climate economics can be derived from professional academic responsibilities. Professional academic responsibility (A3), for example, calls economists to participate in critical debates and to respond to criticism. As presented in Chap. 4, neoclassical climate economics can be criticised from various perspectives and many reasons. These arguments of critics ask for serious responses, as it is common practice in the scientific communities. Our point here is that responsibility to participate in critical discussion requires economists to address climate change in a different way and thus indirectly support the argument for economists' responsibility to address climate change.

### 6.4.3   Responsibility Conflicts in Mainstream Climate Economics

Mainstream climate economics conflicts with several professional academic responsibilities. We summarise the arguments that we presented throughout the book in Table 6.2. Importantly, similar arguments could be raised

**Table 6.2** Responsibility conflicts in mainstream climate economic

| Economists' professional academic responsibilities | Critique on mainstream (climate) economics |
|---|---|
| **(A1) Responsibility to conduct reliable research** | |
| 1.1 Honesty (no misrepresentation of data) | |
| 1.2 Carefulness (minimize errors) | • Uncareful journal-review (Chapter 5.2) |
| 1.3 Avoidance of vested interests | • Vested interests of reviewers (Chapter 5.2)<br>• Selfish pluralism (Chapter 4.2) |
| 1.4 Sharing of data, method, tools, results etc. | |
| 1.5 Obey law & moral standards (e.g. in laboratory) | |
| 1.6 Fair behavior & respect | • Ignoring non-mainstream research (Chapter 4.2)<br>• Feeling of superiority (Chapter 5.3) |
| 1.7 Efficient use of resources and research money | |
| **(A2) Responsibility to teach and educate future scientists** | • Strong standardization in teaching (Chapter 5.5) |
| 2.1 Teaching the whole discipline (variety & depth) | • Lack of pedagogical pluralism (Chapter 5.5)<br>• Economics taught as a textbook science (Chapter 5.5) |
| 2.2 Reflections about Reflections about the philosophy of science, and methodology | • Neglect of history and philosophy of economics (Chapter 5.5) |
| **(A3) Responsibility to participate in academic discussions** | • Lack of interested pluralism (Chapter 4.2) |
| 3.1 about own research, which includes | • Lack of reflexive pluralism |
| 3.1.1 openness (intra-&interdisciplinary) | • Disinterested pluralism (Chapter 4.2) |
| 3.1.2 a response to criticism | • Ignorance of non-mainstream critique (Chapter 5.3)<br>• Avoidance of normative questions (Chapter 4.3) |
| 3.1.3 reflections about the methodology, normativity, philosophy of science | • Avoidance of theory of science (Chapter 5.4)<br>• Methodological weaknesses (Chapter 4.3) |
| 3.2 about research of others, which includes | |
| 3.2.1 to provide feedback, unbiased review | • no free market of ideas (Chapter 5.2) |
| 3.2.2 a general awareness & interest | • lack of disinterested pluralism (Chapter 4.2)<br>• monism, emulation of Physics (Chapter 5.3) |

Source: Authors' own contribution

for the neoclassical treatment of other societal challenges. Neoclassical economics is in a problematic state with respect to the general practice of good science. Besides following the practice of good science, it mainly means to allow for pluralism and an open discussion. In the following, we discuss related responsibility conflicts.

Climate economics is problematic due to its dominance of mainstream neoclassical economics and the resulting lack of pluralism. Instead of interested pluralism, a monist understanding of economics is common practice and implies a lack of openness. However, openness is the key requirement for a constructive exchange between different schools of thought. Openness is not only key for more pluralism in economics but also a key principle for the conduct of good research. According to this principle, openness requires scientists to share results, to allow for discussion and to be open to criticism and new ideas. Especially the latter "prevents science from becoming dogmatic, uncritical, and biased. Openness also contributes to the advancement of science" (Resnik 2005, p. 52). As an example for the missing openness, Negru and Negru (2017, p. 194) point out that none of the neoclassical economists that were invited to the roundtable dialog on pluralism in economics, initiated by the *International Journal of Pluralism and Economics Education* in 2015, accepted the invitation.

In the case of economics, a lack of pluralism and openness lead to several responsibility conflicts especially related to the (A3) *responsibility to participate in academic discussions*. Aspects of (A1.1) fair *behaviour & respect* associated with the (A1) *responsibility to conduct reliable research* also plays a role. Monist economics believe in *one right way* to do research. Consequently, they tend to stay away from discussions about their own and other research, which includes methodological and philosophy of science issues or to go beyond the mainstream boarders. Such refusal of discourse contradicts the (A3) *responsibility to participate in academic discussion* in different ways.

A lack of openness can lead to disinterest in the research of others who follow a different approach. Without knowing what non-mainstream economists are doing, it is also not possible to provide feedback or unbiased review on their work, if asked. Furthermore, mainstream economists who have a monist perspective tend to ignore critiques from heterodox research. They do so with the belief that these researchers are *no proper* researchers, which is why their critique is not worth considering. Consequently, a valid point of critique and thus an opportunity for improvement can be missed. Especially if a feeling of superiority towards heterodox economists exists, problems with mutual respect can arise. Ignoring non-mainstream research on the basis that research that follows a different approach than one's own is *no proper* research contradicts the ideal of respect in research. The ideal demand, however, that "[s]cientists should treat colleagues with respect" (Resnik 2005, p. 60).

As we explained in Sect. 5.5 in Chap. 5, the lack of pluralism and openness in mainstream economics also affects the teaching of economics. We argue that the resulting lack of pluralism in teaching conflicts with the *responsibility to teach and educate future scientists.* This responsibility demands that "[s]cientists should educate prospective scientists and insure that they learn how to conduct good science" (Resnik 2005, p. 56). Furthermore, a well-trained scientist that received a good education has "tacit knowledge of her subject that extends far beyond what she can learn in textbooks or lectures" (Resnik 2005, p. 56). For an understanding of a discipline that goes beyond textbooks, it is necessary to know about the discipline's history, its methodology and the different schools of thought. Thus, dogmatic and textbook style of economic teaching, which leaves little room for reflective thinking, contradicts this ideal in all its aspects. Mearman and McMaster (2019) emphasise the importance of teaching history and philosophy of economics and also ethics to students. These aspects are underrepresented in most curricula, which result in responsibility conflicts. DeMartino (2019) demands a reform of the economics curriculum in order to train students to become ethical economists, which is needed to address societal challenges.

The current dominating practice of teaching economics is not only problematic due to resulting responsibility conflicts. It also has serious implications for the future generation of economists. If the lack of pluralism in economics leads to a lack of pedagogical pluralism in teaching, there is the risk that future economists are also developing monist view about science and rejecting interested pluralism. However, the good practice of science requires openness, which is opposed to monism. When future economists are not trained according to this ideal, they will most likely struggle with the ideal.

As argued in the beginning of this chapter, the crisis of responsibility related to climate change manifests itself in a responsibility denial of different degrees. Section 6.4 shows that not only society but also the science of economic, mainly neoclassical economics, is affected. Mainstream climate economics is missing interested pluralism and openness, which has negative effects on the fulfilment of economists' responsibilities. It is hard to deny that mainstream climate economics lags behind to fulfil their special responsibility—the responsibility to adequately address climate change. The identified responsibility conflicts also point to a crisis concerning the good practice of science of economics that go far beyond climate change.

## REFERENCES

AAAS. 1980. Principles of Scientific Freedom and Responsibility. Washington DC.

Ackerman, Frank. 2008. Climate Economics in Four Easy Pieces. *Development* 51 (3): 325–331.

———. 2013. *Climate Economics:* Routledge.

Agazzi, Evandro. 2014. *Scientific Objectivity and Its Contexts.* Cham: Springer International Publishing.

Akerlof, George, and et al. 2019. Economists' Statement on Carbon Dividends: Bipartisan Agreement on How to Combat Climate Change. *The Wall Street Journal,* January 17.

Alai, Mario. 2016. The Issue of Scientific Realism. In *Science between Truth and Ethical Responsibility: Evandro Agazzi in the Contemporary Scientific and Philosophical Debate,* ed. Mario Alai, Marco Buzzoni, and Gino Tarrozi, 45–64. Cham: Springer International Publishing Switzerland.

Beckwith, Jon, and Franklin Huang. 2005. Should We Make a Fuss? A Case for Social Responsibility in Science. *Nature Biotechnology* 23 (12): 1479–1480.

Bird, Stephanie. 2014. Socially responsible science is more than "good science". *Journal of Microbiology & Biology Education* 15 (2): 169–172.

Bridgman, Percy. 1947. Scientists and Social Responsibility. *The Scientific Monthly* 65 (2): 69–72.

Callon, Michel. 1998. *The Laws of the Markets.* Oxford: Blackwell.

Camenisch, Paul. 1983. On Being a Professional, Morally Speaking. In *Moral Responsibility and the Professions,* ed. Bernard Baumrin and B. Freedman, 42–61. New York: Haven Press.

Caney, Simon. 2005. Cosmopolitan Justice, Responsibility, and Global Climate Change. *Leiden Journal of International Law* 18 (4): 747–775.

Cetto, Ana Maria. 2000. *Science for the Twenty-First Century: A New Commitment.* 26th ed. UNESCO: Paris.

Climate Leadership Council. 2019. Economists' Statement | Climate Leadership Council. Accessed November 4, 2019, from https://www.clcouncil.org/economists-statement/.

Cohen, Stanley. 2013. *States of Denial: Knowing about Atrocities and Suffering.* Cambridge: Polity Press.

Colander, David, Michael Goldberg, Armin Haas, Katarina Juselius, Alan Kirman, Thomas Lux, and Brigitte Sloth. 2009. The Financial Crisis and the Systemic Failure of the Economics Profession. *Critical Review* 21 (2–3): 249–267.

Committee on Science, Institute of Medicine, National Academy of Sciences, National Academy of Engineering. 2009. *On being a scientist: A guide to responsible conduct in research*, 3rd edn. Washington, D.C: National Academies Press.

DeMartino, George. 2011. *The Economist's Oath*. Oxford: Oxford University Press.

———. 2013. Professional Economic Ethics: Why Heterodox Economists Should Care. *Economic Thought* 2 (1): 43–53.

———. 2014. A Professional Ethics Code for Economists. *Challenge* 48 (4): 88–104.

———. 2016. "Econogenic Harm": On the Nature of and Responsibility for the Harm Economists Do as They Try to Do Good. In *The Oxford Handbook of Professional Economic Ethics*, ed. George DeMartino and Deirdre McCloskey, 70–98. New York: Oxford University Press.

———. 2019. Training the 'ethical Economist'. In *The Ethical Formation of Economists*, ed. Ioana Negru and Wilfred Dolfsma, 7–23. New York: Routledge.

DeMartino, George, and Deirdre McCloskey, eds. 2016. *The Oxford Handbook of Professional Economic Ethics*. New York: Oxford University Press.

———. 2018. Professional Ethics 101: A Reply to Anne Krueger's Review of The Oxford Handbook of Professional Economic Professional Economic Ethics. *Econ Journal Watch* 15 (1): 4–19.

Dolfsma, Wilfred, and Ioana Negru. 2019. Preface: The Ethical Formation of Economists. In *In: The Ethical Formation of Economists*, ed. Ioana Negru and Wilfred Dolfsma. New York: Routledge.

Doorn, Neelke. 2017. Allocating Responsibility for Environmental Risks: A Comparative Analysis of Examples from Water Governance. *Integrated Environmental Assessment and Management* 13 (2): 371–375.

Douglas, Heather E. 2003. The Moral Responsibilities of Scientists (Tensions between Autonomy and Responsibility). *American Philosophical Quarterly* 19 (4): 59–68.

Dow, Sheila C. 2013. Codes of Ethics for Economists: A Pluralist View. *Economic Thought* 2 (1): 20–29.

Economists for Future—Rethinking Economics. 2019. Accessed November 4, 2019. http://www.rethinkeconomics.org/projects/economists-for-future/.

Edsall, John T. 1975. Scientific Freedom and Responsibility. *Science (New York, N.Y.)* 188 (4189): 687–693.

Evers, Kathinka. 2001. *Standards for Ethics and Responsibility in Science: An analysis and evaluation of their content, background and function*. https://eclass.upatras.gr/modules/document/file.php/PDE1515/EVERS_%20Standards%20for%20Ethics%20and%20Responsibility%20in%20Science%20book.pdf.

Faulhaber, Gerald R., and William J. Baumol. 1988. Economists as Innovators: Practical Products of Theoretical Research. *Journal of Economic Literature* 26 (2): 577–600.

Fourcade, Marion, Etienne Ollion, and Yann Algan. 2015. The Superiority of Economists. *Journal of Economic Perspectives* 29 (1): 89–114.

Friedman, Milton, and Mariyln Friedman. 1953. *Essays in Positive Economics.* University of Chicago Press.

Gardiner, Stephen. 2011. Is No One Responsible for Global Environmental Tragedy? Climate Change as a Challenge to Our Ethical Concepts. In *The Ethics of Global Climate Change,* ed. Denis Gordon Arnold, 38–59. Cambridge: Cambridge University Press.

Glerup, Cecilie, and Maja Horst. 2014. Mapping 'Social Responsibility' in Science. *Journal of Responsible Innovation* 1 (1): 31–50.

Groot, Wim, and Henriette van den Brink. 2019. Economics, Their Role and Influence in the Media. In *The Ethical Formation of Economists,* ed. Ioana Negru and Wilfred Dolfsma, 132–144. New York: Routledge.

Hagedorn, Gregor, et al. 2019. The Concerns of the Young Protesters Are Justified: A Statement by Scientists for Future Concerning the Protests for More Climate Protection. *GAIA—Ecological Perspectives for Science and Society* 28 (2): 79–87.

Hansen, Tom. 2006. Academic and Social Responsibility of Scientist. *Journal on Science and World Affairs* 2 (2): 71–92.

Heise, Arne. 2009. Toxische Wissenschaft?: Zur Verantwortung der Ökonomen für die gegenwärtige Krise. *Wirtschaftsdienst* 89 (12): 842–848.

Hirschman, D., and E. Berman. 2014. Do Economists Make Policies?: On the Political Effects of Economics. *Socio-Economic Review* 12 (4): 779–811.

ICSU. 1999. *Ethics and the Responsibility of Science.* http://www.unesco.org/science/wcs/backgrounds/ethics_uncertainty.htm.

Jamieson, Dale. 1992. Ethics, Public Policy, and Global Warming. *Science, Technology, & Human Values* 17 (2): 139–153.

Jonas, Hans. 1985. *The Imperative of Responsibility: In Search of an Ethics for the Technological Age.* Chicago, IL: The University of Chicago Press.

Juhola, Sirkku. 2019. Responsibility for Climate Change Adaptation. *Wiley Interdisciplinary Reviews: Climate Change* 2 (1): 1–13.

Kitcher, Philip. 1993. *The Advancement of Science: Science Without Legend, Objectivity Without Illusion.* New York: Oxford University Press.

Krueger, Anne. 2017. Review of Oxford Handbook of Professional Economic Ethics by George F. DeMartino and Deidre McCloskey. *Journal of Economic Literature* 55 (1): 209–216.

Lichtenberg, Judith. 2010. Negative Duties, Positive Duties, and the "New Harms". *Ethics* 120 (3): 557–578.

Longino, Helen. 1990. *Science as Social Knowledge: Values and Objectivity in Scientific Inquiry.* Princeton: Princeton University Press.

LSE. 2017. Professor Lord Nicholas Stern. Accessed November 22, 2019. http://personal.lse.ac.uk/sternn/.

Lübbe, Hermann. 1986. Scientific Practice and Responsibility. In *Facts and Values: Philos. Reflections from Western and Non-Western Perspectives,* ed. Marinus C. Doeser, 81–95. Dordrecht: Nijhoff.

MacKenzie, Donald, and Yuval Millo. 2003. Constructing a Market, Performing Theory: The Historical Sociology of a Financial Derivatives Exchange. *American Journal of Sociology* 109 (1): 107–145.

Mearman, Andrew, and Robert McMaster. 2019. Teaching Future Economists. In *The Ethical Formation of Economists,* ed. Ioana Negru and Wilfred Dolfsma, 24–43. New York: Routledge.

Morgan, Jamie. 2019. Intervention, Policy and Responsibility: Economics as Over-Engineered Expertise? In *The Ethical Formation of Economists,* ed. Ioana Negru and Wilfred Dolfsma, 145–163. New York: Routledge.

Negru, Ioana, and Anca Negru. 2017. Modes of Pluralism: Critical Commentary on Roundtable Dialogue on Pluralism. *International Journal of Pluralism and Economics Education* 8 (2): 193–209.

Neubauer, Luisa-Marie, and Alexander Repenning. 2019. *Vom Ende der Klimakrise: Eine Geschichte unserer Zukunft.* Stuttgart: Tropen.

Norgaard, Kari Marie. 2006. "People Want to Protect Themselves a Little Bit": Emotions, Denial, and Social Movement Nonparticipation*. *Sociological Inquiry* 76 (3): 372–396.

O'Sullivan, Patrick. 2019. Economists' Personal Responsibility and Ethics. In *The Ethical Formation of Economists,* ed. Ioana Negru and Wilfred Dolfsma, 44–60. New York: Routledge.

Oswald, Andrew, and Nicholas Stern. 2019. *Why does the economics of climate change matter so much, and why has the engagement of economists been so weak?* https://www.res.org.uk/resources-page/october-2019-newsletter-why-does-the-economics-of-climate-change-matter-so-much-and-why-has-the-engage-ment-ofeconomists-been-so-weak.html.

Passel, Peter. 1997. Yawn. A Global Warming Alert. But This One Has Solutions. *New York Times,* February 13.

Reif, Alexander, and Cornelius Dahm. 2017. Globale Klimakrise: Aufbruch in eine neue Zukunft. Ursachen, Auswirkungen und transformative Wege aus der Klimakrise. https://germanwatch.org/sites/germanwatch.org/files/GW-Klimakrise-WEB_0.pdf. Accessed 7 November 2019.

Resnik, David. 1996. Social Epistemology and the Ethics of Research. *Studies in History and Philosophy of Science Part A* 27 (4): 565–586.

———. 2005. *The Ethics of Science: An Introduction.* London: Routledge.

Ripple, William Christopher Wolf, Thomas Newsome, Phoebe Barnard, and William Moomaw. 2020. World Scientists' Warning of a Climate Emergency. *BioScience* 5: 8–12.

Roos, Michael. 2016. What Is the Social Responsibility of Academic Economists? *Paper presented at the Conference "40 years of the Cambridge Journal of Economics"*, Cambridge, July 12–13.

Russell, Bertrand. 1960. The Social Responsibilities of Scientists. *Science (New York, N.Y.)* 131 (3398): 391–392.

Sakharov, Andrei. 1981. The responsibility of scientists. *Nature* 291 (5812): 184–185.

Scavenius, Theresa. 2018. Climate Change and Moral Excuse: The Difficulty of Assigning Responsibility to Individuals. *Journal of Agricultural and Environmental Ethics* 31 (1): 1–15.

Schneider, Friedrich, and Gebhard Kirchgässner. 2009. Financial and World Economic Crisis: What Did Economists Contribute? *Public Choice* 140 (3–4): 319–327.

Schneidewind, Uwe, Mandy Singer-Brodowski, Karoline Augenstein, and Franziska Stelzer. 2016. Pledge for a Transformative Science: A conceptual framework. Wuppertal Paper 2016 (191).

Scientists for Future. 2019. Charta von Scientists for Future. Accessed November 11, 2019. https://www.scientists4future.org/about/charta/.

Shrader-Frechette, Kristin. 1994. *Ethics of Scientific Research*. Lanham, MD: Rowman & Littlefield.

Sinnott-Armstrong, Walter. 2010. It's Not My Fault: Global Warming and Individual Moral Obligations. In *Climate Ethics: Essential Readings*, ed. Stephen Mark Gardiner, 332–346. Oxford and New York: Oxford University Press.

Sombetzki, Janina. 2014. *Verantwortung als Begriff, Fähigkeit, Aufgabe: Eine Drei-Ebenen-Analyse*. Wiesbaden: Springer.

Steigleder, Klaus. 2018. Climate Economics and Future Generations. In *Towards the Ethics of a Green Future: The Theory and Practice of Human Rights for Future People*, ed. Marcus Düwell, Gerhard Bos, and Naomi van Steenbergen, 131–153. London and New York: Routledge.

Stern, Nicholas. 2007. *The Economics of Climate Change: The Stern Review*. Cambridge: Cambridge University Press.

———. 2011. *How should we think about the economics of climate change. Lecture for the Leontief Prize*. Medord. http://www.ase.tufts.edu/gdae/about_us/leontief/SternLecture.pdf.

Stilgoe, Jack, Richard Owen, and Phil Macnaghten. 2013. Developing a Framework for Responsible Innovation. *Research Policy* 42 (9): 1568–1580.

Strohschneider, Peter. 2014. Zur Politik der Transformativen Wissenschaft. In *Die Verfassung des Politischen*, ed. André Brodocz, Dietrich Herrmann, Rainer Schmidt, Daniel Schulz, and Julia Schulze Wessel, 175–192. Wiesbaden: Springer.

UNFCCC. 2015. Paris Agreement on Climate Change.

Verhoog, Henk. 1981. The Responsibilities of Scientists. *Minerva* 19 (4): 582–604.

Young, Iris Marion. 2011. *Responsibility for Justice*. New York: Oxford University Press.

Ziman, John. 1998. Why Must Scientists Become More Ethically Sensitive Than They Used to Be? *Science (New York, N.Y.)* 282 (5395): 1813–1814.

# CHAPTER 7

# Concluding Thoughts

**Abstract** In the conclusion, the main insights are summarised. Roos and Hoffart explain why the problems related to neoclassical economics discussed in this book are not only relevant for climate change. These problems are of importance for so-called wicked problems and other societal challenges, too, such as the loss of biodiversity, inequality and social polarisation, and the design and regulation of financial markets. In all these cases, criticisms similar to the ones the authors raised against neoclassical climate economics apply. The book ends with some thoughts on what could be done to transform mainstream economics into a more plural and responsible discipline.

**Keywords** Failure of economics • Societal challenge • Interested pluralism • Discussion • Transformation of economics

Economics is in a bad state. It is dominated by the neoclassical school of thought with its particular perspective on the world and on how to do research. Economics is detached from other social sciences and from many problems that bother society. We argued that current mainstream economics has little to say about climate change, which is one of the most pressing challenges of our time.

157
M. Roos, F. M. Hoffart, *Climate Economics*, Palgrave Studies in Sustainability, Environment and Macroeconomics, https://doi.org/10.1007/978-3-030-48423-1_7

Defenders of mainstream economics might respond that this accusation is not justified. As exonerating evidence they might present the Nobel Prize in economics awarded to William D. Nordhaus for "integrating climate change into long-run macroeconomic analysis" in 2018. Pointing to the Nobel Prize can serve two purposes. First, it suggests that there is research of highest quality in economics that deals with climate change and that such research pays off scientifically. Second, since Alfred Nobel wanted research to be prized, that is, "for the greatest benefit to humankind", one cannot claim that economic research is not relevant to society.

This exonerating evidence is weak. The Nobel Prize may be awarded for the best research according to mainstream criteria of economics, but this does not imply that these criteria are good and that the research could not be better. Nordhaus' attempt to determine the optimal climate mitigation policy is unlikely to lead to reliable answers. The search for the right social cost of carbon is not only a waste of scientific resources but also harmful to society. In many versions of the DICE model, the optimal carbon price would lead to a rise in atmospheric temperature well beyond 2°C, which many natural scientists consider to be highly risky for ecosystems. The DICE models trade damages to ecosystems off against economic benefits. Keeping temperature rise below 2 °C by effective mitigation measures would reduce economic growth in a supposedly suboptimal way. This is dangerous thinking. Suppose that the entire global agricultural systems collapsed due to the damages to ecosystems. Most people would consider this as a serious problem to humankind, but Nordhaus' logic says that the problem was rather small, since agriculture, forestry and fishing contribute less than 4% to global GDP (Hickel 2018). Nordhaus conveys the impression that there is a scientific basis for slow climate change mitigation that affects economic growth in the least possible way. The Nobel Prize lends public support to this impression. Politically, Nordhaus' message is attractive, because it is difficult to organise political majorities for large economic transformations. Policymakers seem to have a choice whether to follow the gloomy advice of natural scientists to cut $CO_2$ emissions harshly at large economic costs or the seemingly more attractive message of economists that a less painful path is possible and even advisable. Hickel (2018) points to a coincidence that symbolises the difference between economics and science:

> Strangely, in what seems a bizarre coincidence, Nordhaus was announced as the winner of the Nobel Prize on the very same October day that the IPCC

published its latest report on climate change. The report is the United Nations' most urgent yet: It calls for the world to cut emissions in half by 2030, and get to net zero by the middle of the century. While Nordhaus has spent most of the past four decades calling for gradualism to preserve the conditions for economic growth, the IPCC calls for radical and immediate action in order to preserve the conditions for life. Growth versus life. The conflict between economics and science has never been clearer.

What has been said in this book does not only hold for how economics deals with climate change but also for other great societal challenges. The failure of neoclassical thinking to address climate change in a comprehensive and constructive way is just an example of a more general failure of the economic mainstream to do responsible research.

Although the loss of biodiversity is also linked to economic activity, it is even less on the economic research agenda than climate change. The global extinction of plant and animal species is estimated to be 100 to 1000 times higher than its normal rate (Ceballos et al. 2015; Vos et al. 2015). Biodiversity loss is one of the nine *planetary boundaries* that has been transgressed most clearly and hence is a risk for humanity at least as great as climate change (Cardinale et al. 2012). According to Rockström et al. (2009, p. 32), the planetary boundaries demarcate a safe space within which

> humanity can operate safely. Transgressing one or more planetary boundaries may be deleterious or even catastrophic due to the risk of crossing thresholds that will trigger non-linear, abrupt environmental change within continental- to planetary-scale systems.

Another great societal challenge on which economics has been silent for a long time is social inequality and polarisation. Labour market polarisation grew in many countries as a consequence of globalisation. Autor et al. (2016) show that China's integration into global trade in the 1990s had significant negative impacts on many local labour markets in the U.S. Paul Krugman argues[1] that there was a consensus among mainstream trade economists in the 1990s that globalisation would have some adverse impact on low-skilled workers in advanced countries, but that these effects would be relatively small compared to the overall benefits of more trade.

---

[1] For example, https://www.bloomberg.com/opinion/articles/2019-10-10/inequality-globalization-and-the-missteps-of-1990s-economics.

It is now obvious that this consensus was wrong and that economists underestimated how large and persistent the effects on local labour markets would be in terms of unemployment and wage reductions. Krugman links the political discontent of voter for populists to the negative effects of globalisation. Remarkably, most economists are not particularly worried about digitisation, although it is likely to lead to another wave of labour market polarisation. The consensus narrative is that there were always luddites afraid of new technologies who were proven wrong by history. In the end, at least as many new jobs will be created than will be lost due to the new technology. While this argument might be right in the aggregate and in the long run, it may have the same flaw as the prediction that globalisation will have positive net benefits. Mainstream economics has a history of neglecting the issue of inequality and its impact on society (Wade 2012; Cook 2018). It was a sensation that Thomas Piketty's book on inequality received so much public and scholarly attention. This "Pikettymania" (Cook 2018, p. 20) is a symptom of how unusual it is that eminent economists pay attention to inequality.

Finally, the financial and economic crisis of 2007–2009 revealed another blind spot of mainstream economics. In their famous letter to the Queen, the economists of the British Academy explained[2]:

> So in summary, Your Majesty, the failure to foresee the timing, extent and severity of the crisis and to head it off, while it had many causes, was principally a failure of the collective imagination of many bright people, both in this country and internationally, to understand the risks to the system as a whole.

The monist attitude of the economic mainstream is a main cause of "failure of collective imagination" of economists. The world is complex and surprising such that using only one filter to look at it necessarily leads to blind spots. The combination of blind spots, strong (ideological) beliefs in markets, great confidence in one's analytic and predictive abilities and considerable political influence is a dangerous contribution to the public debate about societal challenges.

Most great societal challenges are so-called *wicked problems* (Rittel and Webber 1973) that cannot be "solved" in the technical sense that the solution eliminates the problem. Instead the "solution" can only improve the

---

[2] http://wwwf.imperial.ac.uk/~bin06/M3A22/queen-lse.pdf.

situation in some dimensions of the problem. Wicked problems are characterised by different frames for defining the problem that depend on the world views of the different stakeholders. The quality of a solution cannot be assessed objectively, but depends on the stakeholders' values and goals. These properties of wicked problems pose a major challenge not only to public policy, but also to traditional science. One way to deal with wicked problems is the collaborative strategy (Roberts 2000), by which all stakeholders try to reach a consensus on how to define the problem, how a generally acceptable solution might look like and how it could be reached. Such consensus should be reached by scientific and societal discourse which is closely connected to the acknowledgement of pluralism of world views and values. As long as economics remains a monist discipline, economists cannot engage in a constructive discourse, because a discourse presupposes that each participant respects the rights of other participants. Currently, this is not the case in mainstream economics due to the systematic vilification of heterodox economists as inferior researchers.

Economists have a responsibility to contribute constructively to the "solution" of societal challenges, such as climate change, which is for Levin et al. (2012) even a "super wicked" problem. The most fundamental level of the responsibility is to establish a practice of true discourse within the discipline itself, for which *interested pluralism* is mandatory. How could economics transform itself into a discipline that welcomes and practises interested pluralism?

The transformation of economics requires changes both at the structural and the individual level. Structural changes concern how research and teaching are organised. In terms of research, it is necessary to reduce the power of journal editors and reviewers, while, at the same time, maintaining quality standards. The peer-review process is surely not perfect, but it does serve the purpose of necessary quality control. Some filtering of research publications is necessary, because otherwise researchers would drown in the sheer number of publications. Open review or open assessment journals, such as the e-journal *Economics* (http://www.economics-ejournal.org/), seem promising, but currently lack prestige. It might help if major research institutions endorsed and incentivised publications in such journals. Open review means that the peer review process is made public and subject to comments by everyone. Sometimes the review process is not even anonymous. Making the review process transparent could foster an open scientific discourse and reduce the power of researchers with strong (ideological) beliefs on how science should be. Furthermore, hiring and funding policies should be reformed in order to foster pluralism. Rethinking economic teaching is important, for instance, by incorporating history of economic thought and economic methodology into the curriculum. New textbooks with a pluralist perspective would help, too, but ideally economics would move at least somewhat away from canonical textbooks and turn to original sources. Reading and discussing original texts is standard in the humanities and other social sciences and would help economists to train their discursive capabilities. Introducing "ethical training" into the curriculum of economists is necessary, too (Mearman and McMaster 2019). In addition to discursive skill, ethical training would entail the acquisition of practical wisdom and self-awareness (DeMartino 2019).

Stating what should be done does not answer the question of how it might be achieved. As with any system transformation, a transition from one state to another results from a combination of external and internal forces. External and internal stakeholders of economics must exert pressure on the discipline to induce a systemic transformation. Society as a whole can be seen as an external stakeholder, which is represented by the government and non-governmental organisations. They can have an impact on science, because they provide the funding for research. Major research funding programmes, such as the EU's *Horizon 2020* and its successor *Horizon Europe*, explicitly require researchers to address policy priorities and societal concerns. In the long run, it can be expected that research activities will respond to these financial incentives. An alternative to financial incentivisation would be direct government intervention into the scientific system by policymakers, for example, by regulations regarding the appointments of researchers at research institutions or regulations concerning the teaching curricula at universities. Academic freedom is an important value that should not be constrained without good reasons. But even a neoclassical logic of market failure would suggest that a direct government intervention can be necessary in some cases. Students are important stakeholders, too, who are situated at the border of the scientific system. They are not full members of academia, yet, but trained by the system and constitute the pool of future researchers. The ongoing pressure of students worldwide to reform economic teaching is a powerful driver of change.

The institutional framework does not change by itself and it is not exogenous to the scientific community. We need economists with a strong sense of professional and social responsibility who question dominant conventions and quality standards in the scientific community. It is not necessary that everybody steps outside the community and ruins his or her career by actively rebelling against the mainstream—though it might help if more people did. But every researcher can look for her niche and the small things that can be done. Every academic economist should always wonder whether an argument is based on logic and evidence or convention, authority and ideology, whether one's own research or the work submitted or presented by somebody else is relevant or just cute, and whether some statement about economic policy is based on science or on private interests. Whenever an argument or a policy statement is based on conventions, opinions, private interests or just irrelevant, serious scholars should voice disagreement. Ultimately, the responsibility of academic economists is to behave like critical and earnest scientists.

Is it realistic to expect that economics will change fast enough to make a significant contribution to tackle climate change? According to Max Planck's famous quip, science advances one funeral at a time. If this were true, economics might come too late to have a serious impact on climate change mitigation, because the time remaining for effective mitigation policies is scarce. Given that the major financial crisis ten years ago did not cause a major revolution in economics, one can indeed believe that it takes a generation for a paradigm shift to occur. Yet, external pressures on economics are rising and might become even stronger with the effects of climate change becoming more visible in the upcoming years. As we know from complexity science, systems can look quite robust and stable for a long time, but slide into a different state quite suddenly once a bifurcation point is passed. Maybe many small internal and external stimuli are moving the system closer and closer to such a tipping point. In the 1970s, we saw Keynesian macroeconomics collapsing rather quickly when the empirical phenomenon of stagflation appeared, which was not predicted by the existing models. But the theoretical edifice collapsed only, because its foundations were already undermined by the rational expectations model that was published by Muth (1961). This experience suggests that change will come if an external shock hits a system that has already been destabilised by internal dynamics. It is, hence, important to work out theoretical and methodological alternatives to mainstream climate economics to be prepared for the next unavoidable external shock.

# REFERENCES

Autor, David H., David Dorn, and Gordon H. Hanson. 2016. The China Shock: Learning from Labor-Market Adjustment to Large Changes in Trade. *Annual Review of Economics* 8 (1): 205–240.

Cardinale, Bradley J., J. Emmett Duffy, Andrew Gonzalez, David U. Hooper, Charles Perrings, Patrick Venail, Anita Narwani, et al. 2012. Biodiversity Loss and Its Impact on Humanity. *Nature* 486 (7401): 59–67.

Ceballos, Gerardo, Paul R. Ehrlich, Anthony D. Barnosky, Andrés García, Robert M. Pringle, and Todd M. Palmer. 2015. Accelerated Modern Human-Induced Species Losses: Entering the Sixth Mass Extinction. *Science Advances* 1 (5). https://doi.org/10.1126/sciadv.1400253.

Cook, Eli. 2018. The Great Marginalization: Why Twentieth Century Economists Neglected Inequality. *Real-World Economics Review 83*: 20–34.

DeMartino, George F. 2019. Training the 'Ethical Economist'. In *The Ethical Formation of Economists*, ed. Ioana Negru and Wilfred Dolfsma, 7–23. New York: Routledge.

Hickel, Jason. 2018. The Nobel Prize for Climate Catastrophe. *Foreign Policy*. https://foreignpolicy.com/2018/12/06/the-nobel-prize-for-climate-catastrophe/.

Levin, Kelly, Benjamin Cashore, Steven Bernstein, and Graeme Auld. 2012. Overcoming the Tragedy of Super Wicked Problems: Constraining Our Future Selves to Ameliorate Global Climate Change. *Policy Sciences* 45 (2): 123–152.

Mearman, Andrew, and Robert McMaster. 2019. Teaching Future Economists. In *The Ethical Formation of Economists*, ed. Ioana Negru and Wilfred Dolfsma, 24–43. New York: Routledge.

Muth, John F. 1961. Rational Expectations and the Theory of Price Movements. *Econometrica* 29 (3): 315.

Rittel, Horst W.J., and Melvin M. Webber. 1973. Dilemmas in a General Theory of Planning. *Policy Sciences* 4 (2): 155–169.

Roberts, Nancy C. 2000. Wicked Problems and Network Approaches to Resolution. *International Public Management Review* 1 (1): 1–19.

Rockström, Johan, Steffen Will, Kevin Noone, Asa Persson, et al. 2009. Planetary Boundaries: Exploring the Safe Operating Space for Humanity. *Ecology and Society* 14 (2): 32.

de Vos, Jurriaan M., Lucas N. Joppa, John L. Gittleman, Patrick R. Stephens, and Stuart L. Pimm. 2015. Estimating the Normal Background Rate of Species Extinction. *Conservation Biology: The Journal of the Society for Conservation Biology* 29 (2): 452–462.

Wade, Robert. 2012. Why Has Income Inequality Remained on the Sidelines of Public Policy for So Long? *Challenges* 55 (3): 21–50.

# INDEX[1]

**A**

Abatement function, *see* Function
Activism, 2, 10, 137
Agent
  average, 78, 81
  heterogeneous, 64, 81
  interdependent, 85
  representative, 52, 78, 81
*American Economic Journal–Applied Economics*, 24
*American Economic Review* (AER), 101, 102, 14, 22, 23, 23n4
Austrian Economics, 62, 63, 66, 73

**B**

Barbarization, 88
Behavioural economics/psychological economics
  new, 63, 64
  old, 63

**C**

*Cambridge Journal of Economics*, 100
Capitalism, 68, 69, 90–91
Capital stock, 42, 53, 73
Carbon price/CO2 pricing/carbon pricing
  optimal, 47, 59, 158
  right, 59, 84, 125
Carbon tax, 7, 8, 36, 37, 46, 50, 87, 91, 140
Climate change
  adaptation, 122, 125
  anthropogenic, 2, 3, 125, 127
  denial, 2, 122, 126–128, 149
  mitigation, 7, 10, 122, 123, 125, 126, 131, 158, 163
Climate economics/climate change economics
  critique on, 58, 70–83, 121, 122
  heterodox, 60n1

[1] Note: Page numbers followed by 'n' refer to notes.

© The Author(s) 2021
M. Roos, F. M. Hoffart, *Climate Economics*, Palgrave Studies in Sustainability, Environment and Macroeconomics,
https://doi.org/10.1007/978-3-030-48423-1